Big Data Analytics
A Social Network Approach

Editors

Mrutyunjaya Panda
Computer Science Department
Utkal University
Bhubaneswar, India

Aboul-Ella Hassanien
University of Cairo
Cairo, Egypt

Ajith Abraham
Director, Machine Intelligence Research Labs
Auburn, Washington, USA

CRC Press
Taylor & Francis Group
Boca Raton London New York

CRC Press is an imprint of the
Taylor & Francis Group, an **informa** business

A SCIENCE PUBLISHERS BOOK

CRC Press
Taylor & Francis Group
6000 Broken Sound Parkway NW, Suite 300
Boca Raton, FL 33487-2742

© 2019 by Taylor & Francis Group, LLC
CRC Press is an imprint of Taylor & Francis Group, an Informa business

No claim to original U.S. Government works

Version Date: 20180803

International Standard Book Number-13: 978-1-138-08216-8 (Hardback)

Library of Congress Cataloging-in-Publication Data

Names: Panda, Mrutyunjaya, editor. | Abraham, Ajith, 1968- editor. | Hassanien, Aboul Ella, editor.
Title: Big data analytics. A social network approach / editors, Mrutyunjaya Panda, Computer Science Department, Utkal University, Bhubaneswar, India, Ajith Abraham, Director, Machine Intelligence Research Labs, Auburn, Washington, USA, Aboul-Ella Hassanien, University of Cairo, Cairo, Egypt.
Description: Boca Raton, FL : Taylor & Francis Group, [2018] | "A science publishers book." | Includes bibliographical references and index.
Identifiers: LCCN 2018029675 | ISBN 9781138082168 (hardback : acid-free paper)
Subjects: LCSH: Big data. | Discourse analysis, Narrative. | Truthfulness and falsehood. | Online social networks.
Classification: LCC QA76.9.B45 B547 2018 | DDC 005.7--dc23
LC record available at https://lccn.loc.gov/2018029675

Visit the Taylor & Francis Web site at
http://www.taylorandfrancis.com

and the CRC Press Web site at
http://www.crcpress.com

Preface

The popularity of social networking has dramatically increased in recent years, which has led to a rise in the amount of data created daily. Due to this huge amount of data, the traditional software algorithm is no longer efficient and big data analysis is required. At the same time, there is the risk of malicious users attacking others through their social relationships. This poses a big challenge. Based on the diversity and complexity of social network research in this area of big data, this book will address social network mining as well as required security measures.

The book contains 14 chapters, which are self contained, presented in a simple manner in order to create interest among readers with greater flexibility.

We are very grateful to the authors of this volume and to the reviewers for their excellent cooperation by critically reviewing the chapters. Most of the authors of the chapters included in this book also served as referees to review the work of their peers. We extend our sincere thanks to CRC Press for their editorial assistance in publishing this edited volume. We hope this edited volume shall serve as an excellent material for the readers who work in this field and will attract subsequent advancement in this subject.

July, 2018

Mrutyunjaya Panda
Utkal University, Bhubaneswar, India

Aboul-Ella Hassanien
Cairo University, Egypt

Ajith Abraham
MIR Labs, Auburn, WA USA

Content

CHAPTER **1**

Linkage-based Social Network Analysis in Distributed Platform

Ranjan Kumar Behera,[1,*] *Monalisa Jena,*[2] *Debadatta Naik,*[1]
Bibudatta Sahoo[1] *and Santanu Kumar Rath*[1]

Introduction

The social network is a platform where a large number of social entities can communicate with each other by sharing their views on a number of topics, posting a bunch of multimedia files or exchanging a number of messages. It is the structure, where the group of users is connected through their social relationships. The social network can be modeled as a non-linear data structure, like the graph, where each node represents a user and the edges between them depict the relationships between the users. Facebook, Twitter, YouTube, LinkedIn, Wikipedia are some of the most popular social network platforms where billions of users interact every day. In these social network websites, along with the interaction, exabyte of structured, unstructured and semi-structured data is generated at every instance of time (Hey et al. 2009). This is the reason where the term 'big data' is associated with the social network. Research in the social network has the beginning with the social scientist analyzing human social behavior,

---

[1] National Institute of Technology, Rourkela, 769008, India.
[2] Fakir Mohan University, Balasore, 756019, India.
Emails: bmonalisa.26@gmail.com; deba.uce03@gmail.com; bdsahu@nitrkl.ac.in;
skrath@nitrkl.ac.in
* Corresponding author: jranjanb.19@gmail.com

mathematicians analyzing complex network theory and now computer scientists analysing the generated data for extracting quite a number of useful information. Social scientists, physicists, and mathematicians are basically dealing with the structural analysis of social network while computer scientists are dealing with data analysis. Social networks are the most important sources of data analytics, assessment, sentiment analysis, online interactions and content sharing (Pang et al. 2008). At the beginning stage of a social network, information was posted on the homepages and only a few internet users were able to interact through the homepages. However, nowadays an unimaginable number of activities are carried out through the social network which leads to a huge amount of data deposition. Social network enables the users to exchange messages within a fraction of time, regardless of their geographical location. Many individuals, organizations and even government officials now follow the social network structure and media data to extract useful information for their benefit. Since the data generated from the social network are huge, complex and heterogeneous in nature, it proves a highly computationally challenging task. However, big data technology allows analysts to sift accurate and useful information from the vast amount of data.

Social network analysis is one of the emerging research areas in the era of big data analytics. Social network consists of linkage and content data (Knoke and Yang 2008). Linkage data can be modeled through graph structure which depicts the relationship between the nodes whereas the content data is in the form of structured, semi-structured or unstructured data (Aggarwal 2011). They basically consist of text, images, numerical data, video, audio, etc. Basically social network analysis can be broadly classified into two categories: Social Media Analysis (He et al. 2013) and Social Structure Analysis (Gulati 1995).

It has been observed that a huge amount of data is generated from social network at every fraction of time and the size of generated data is increasing at an exponential rate. Storing, processing, analyzing the huge, complex and heterogeneous data is one of the most challenging tasks in the field of data science. All the data which is being created on the social network website can be considered as social media. The availability of massive amounts of online media data provides a new impetus to statistical and analytical study in the field of data science. It also leads to several directions of research where statistical and computational power play a major role in analyzing the large-scale data. The structure-based analysis is found to be more challenging as it is a more complex structure than the media data. A number of real-time applications are based on structural analysis of the network where linkage information in the network plays a vital role in analysis. Social network is the network of relationships between the nodes where each node corresponds to the user and the link between them corresponds to the interaction between them. The interaction may be friendship, kinship, liking, collaboration, co-authorship or any other kind of relationship (Zhang et al. 2007). The basic idea behind each social network is the information network where groups of users either post and share common information or exchange information between them. The concept of social network is not restricted to particular types of information network; it could be any kind of network between the social entities where information is generated continuously. A number analysis can be carried out using the structural information of the network to identify the importance of each node or reveal the hidden relationships between the users. Before discussing the details of research in social network analysis, it is better to point out certain kinds of structural properties that real-world social network follows. Some of them are small-world phenomenon, power-law degree distribution or scale-free network, etc., which were devised much before the advent of computer and internet technology. Small-world phenomenon was proposed by Jeffery and Stanley Milgram in 1967 which says that most of the people in the world are connected through a small number of acquaintances which further leads to a theory known as six-degree of separation (Kleinfeld 2002, Shu and Chuang 2011). According to the theory of six-degree of separation, any pair of actor in the planet are separated by at most six degree of acquaintances. This theory is now the inherent principle of today's large-scale social network (Watts and Strogatz 1998). A number of experiments were carried out by social scientists to prove the six-degree of separation principle. One of the experiments is reflected in MSN messenger data. It shows that the average path length between any two MSN messenger users is 6.6. The real-world social network is observed to follow power-law degree distribution, which implies that most of the nodes in the network are having less degree and few of the nodes are having a larger degree. The fraction of nodes having k connection with other nodes in the network depends on the value of k and a constant parameter. It can be mathematically defined as follows (Barabsi and Albert 1999):

$$F(k) = k^{-\gamma} \tag{1}$$

where F(k) is the fraction of node having out-degree k and y is the scale parameter typically ranging from 2 to 3. Scale-free network is the network which follows the power-law degree distribution. We can say that the real-world social network is scale-free in nature rather than following a random network where degree distribution among the nodes is random. Traditional tools are inefficient in handling a huge amount of unstructured data that are generated from the large-scale social network. Apart from the generated online social media data, the structure of the social network is quite complex in nature, being difficult to analyze. A distributed platform like Spark (Gonzalez et al. 2014), Hadoop (White 2012) may be a suitable platform for analyzing large-scale social network efficiently.

Research Issues in Social Network

Nowadays the social network is observed to be an inevitable part of human life. The fundamental aspect of social networks is that they are rich in content and linkage data. The size of the social network is increasing rapidly as millions of users along with their relationships are added dynamically at every instance of time (Borgatti and Everett 2006). A huge amount of data is generated exponentially through the social network. The social network analysis is based on either linkage data or the content data of the network. A number of data mining and artificial intelligence techniques can be applied to these data for extracting useful pattern and knowledge. As the size of the data is huge, complex and heterogeneous in nature, traditional tools are inefficient in handling such data. Big data techniques may be helpful in analyzing the ever-increasing content data of the social network. However, in this chapter, we mainly focus on the structural analysis of the social network. Structural analysis can be carried out on either static or dynamic network. Static analysis of the network is possible only when the structure of the network does not change frequently, like in bibliographic network or co-authorship network where the author collaboration or citation count increases slowly over time. In static analysis, the entire network is processed in batch mode. Static analysis is easier in comparison to dynamic analysis where the structure of the networks is changing at every instance of time.

A large number of research problem may evolve in the context of structural analysis of the network. As we know, the structure of the social network changes at an exponential rate in the era of the Internet, the attributes involved in dynamics of the network also change. Modeling the dynamic structure of the social network is found to be quite a challenging task as the network parameters are changing more rapidly than expected. For example, as per small-world phenomena, any two entities are supposed to be separated by a small number of acquaintances but the actuality of this phenomenon may be fickle over the structural changes of the network. Verifying several structural properties in the dynamic network is found to be of great interest in recent years.

Linkage-based Social Network Analysis

Centrality Analysis

Centrality analysis is the process of identifying the important or most influential node in the network (Borgatti and Everett 2006). The meaning of importance may differ from application to application. In one context, a specified node is found to be the most influential node while in another context, the same node might not have a higher influence factor in the network, for example, in a social network website, it can be inferred that at one instance of time, Obama is the most powerful person in the world while in another instance of time, he might not be that powerful. Centrality analysis could be helpful in finding the relative importance of the person in the network. The word importance has a wide number of meanings that lead to different kinds of centrality measures. Evaluating centrality values for each node in a complex network is found to be quite a challenging task and considered as one of the important applications in the large-scale social network where data size is large. The value of centrality of a node might differ in the variety of centrality measure. The importance of the nodes can be analyzed by the following centrality measures:

Degree centrality

Degree centrality is the simplest form of centrality measure which counts the number of direct relationships of a node with all nodes in the network. The generalization of the degree centrality can be k-degree centrality, where k is the length of the path or connection from one node to all other nodes. The intuition behind this centrality measure is that people with a greater number of relationships or connections tend to be more powerful in the network. Measuring the degree centrality in the small network may be an easier task, where it is a computationally challenging for a large-scale network with millions of nodes connected with billions of edges. It can be mathematically represented as follows (Freeman 1978):

$$DC(i) = \frac{\sum_{j=1}^{n} a_{ij}}{n-1} \qquad a_{ij} = \begin{cases} 1 & \text{if } (i,j) \in E \\ 0 & \text{if } (i,j) \notin E \end{cases} \qquad (2)$$

where $DC(i)$ is the degree centrality of the node i; n is the total number of nodes in the network; a_{ij} represents the element of adjacency matrix for the network; the value of a_{ij} is 1, if there is an edge between node i and node j; otherwise, its value is 0. Degree centrality can be extended to the network level. It can be used to measure the heterogeneity of the network.

The measure of the degree variation in the network can be specified by the degree of centrality of the network which can be mathematically defined as follows:

$$DC(network) = \frac{\sum_{i=1}^{n} [DC(\text{max}) - DC\,(i)]}{(n-1)\,(n-2)} \tag{3}$$

Here, *DC(network)* is the degree of centrality for the network; *DC(max)* is the maximum degree of the network.

Betweenness centrality

Betweenness centrality of a node is used to measure the amount of flow information that passes through the node. It is defined as the number of the shortest path between any pair of nodes that passes through the node in the network (Barthelemy 2004, Newman 2005). It is the most useful centrality measure that has a number of real-time applications. For example, in telecommunication cation network, a node with high betweenness centrality value may have better control over information flow because more information can pass through the node in community detection. Node with higher betweenness value may act as the cut node between two communities, in software-defined network. It can be used to find the appropriate position of the controller. Betweenness centrality of a node can be mathematically defined as follows (Brandes 2001):

$$BC(i) = \sum_{j \neq i \neq k} \frac{nsp_{jk}(i)}{nsp_{jk}} \tag{4}$$

where *BC(i)* is the betweenness centrality for the node *i*; $nsp_{jk}(i)$ is the number of shortest paths between node *j* and *k* and which pass through node *i*; and nsp_{jk} is the total number of shortest paths between *j* and *k*.

Eigen-vector centrality

Eigen-vector centrality of the nodes is based on the principle of liner algebra which has a number of theoretical applications in the social network (Ruhnau 2000). The intuition behind eigenvector centrality is that a node is said to be more central if it is connected to more important nodes rather than having the connection with a greater number of unimportant nodes. According to the Perron Frobenius theorem, values of the eigenvector corresponding to highest eigen value of adjacency matrix represent the centrality value of each node. It can be defined as (Newman 2008):

$$EC(i) = \sum_{j=1}^{n} [a_{ij} * EC(j)] \quad where\; a_{ij} = \begin{cases} 1 & \text{if } (i,j) \in E \\ 0 & \text{if } (i,j) \notin E \end{cases} \tag{5}$$

where $EC(i)$ is the eigenvector centrality of node i; a_{ij} represents the element of the adjacency matrix. From equation 5, it can be observed that eigen centrality of node i depends on the eigen centrality of neighbouring node (j).

Closeness centrality

Closeness centrality is the measure of closeness of a node to all other nodes. A node which is more close to all other nodes may have a high influence factor in the network (Okamoto et al. 2008). In the social network, people who are more close to all other people may have a higher influence on others. Closeness centrality is nothing but the reciprocal of distance of a node from the rest of the nodes. It can be expressed as follows (Rahwan 2016):

$$CC(i) = \frac{1}{\sum_{j=1}^{n} d_{ij}} \tag{6}$$

where $CC(i)$ is the closeness centrality of node i and d_{ij} is the distance between nodes i and j.

Community Detection

Communities are found to be one of the most important features of the large-scale social network. Uncovering such hidden features enables the analysts to explore the functionalities of the social network. There exists quite a good number of definitions of community depending on the contexts pertaining to different applications. However, as per a number of commonly accepted definitions, they are considered to be a group of nodes which have a dense connection among themselves as compared to sparsity outside the group (Tang and Liu 2010, Newman and Girvan 2004). Communities in the social network represent a group of people who share common ideas and knowledge in the network. Their identification helps in getting insight into social and functional behavior in the social network. However, due to unprecedented growth in the size of the social network, it is quite a challenging task to discover subgroups in the network within a specified time limit. A number research in the direction of community detection was done recently. Community detection algorithms were basically classified into the following types (Fortunato 2010, Day and Edelsbrunner 1984):

- Node-based community detection
- Group-based community detection
- Network-based community detection
- Hierarchy-based community detection

In node-based community detection algorithms, group of nodes are said to form a community, if each node in the group satisfies certain network properties, like k-clique, where each node has connection with at least k number other nodes; k-plex where each node in the group has at least n-k number of connections. In group-based community detection algorithm, the group of nodes must meet the required constraints without zooming into the node level. A node within the community may not satisfy the criteria but as a group, if it satisfies the network properties, it can be treated as a community. Network-based community detection algorithm depends on certain criteria which must be satisfied by the network as a whole. Modularity and clustering coefficient are some of the well-known parameters that are used in network-based community detection algorithm. Communities are detected in a recursive manner in hierarchy-based community detection algorithms. It is further divided into two types—one is the agglomerative approach (Day and Edelsbrunner 1984) and the second is the divisive approach (Reichardt and Bornholdt 2006). In agglomerative approach, initially each node is considered as a community. At each step, the set of nodes are grouped into the community in a way that improves certain network parameters, like modularity. Grouping of nodes is considered to be a community at a certain point, where no further improvement of the parameter is possible. Unlike agglomerative approach, divisive algorithms constitute the top-down approach where the whole network is considered as a community at the initial stage. Consequently, nodes are partitioned into communities by removal of edges from the network in a way that improves the network parameter. Similar to the first approach set of nodes is grouping into communities where no further improvement of the parameter is possible.

Community detection in the large-scale network is an extremely challenging task as it involves high computational complexity. In the real-world social network, the structure of the network changes frequently as a number of nodes are added and/or removed at every instance of time. Structure of communities may change over time which the analyst has difficulty in identifying.

Link Prediction

Link prediction is found to be the most interesting research problem in social network analysis. Here the analyst tries to predict the chances of formation of hidden links in the social network. It has a wide number of applications, such as suggesting friends in the online social network, recommending e-commerce product to users, detecting links between terrorists so as to avoid future unwanted circumstances. The basic objective of this problem is to determine the important future linkage information in the network (Liben-Nowell and Kleinberg 2007). By utilizing the linkage information, one can predict the future potential relationships and hidden links in the network. Link prediction in the large-scale network is found to be an NP-complete problem. Most of the researchers are utilizing structural information to make the prediction. However, it may also be possible to predict the link using node attributes. Structural link prediction can be classified as node-based or the path-based. Node-based link prediction is based on the similarity score between the nodes computed by utilizing the local topological structure around the node. A number of similarity indices have been proposed by several researchers in literature. Path-based link prediction score is based on the global structural information of the network. The intuition behind the path-based link prediction is that if two nodes are going to be connected in future, then there must be a path between the two nodes in current topological information. Less is the path length the higher the chance of them getting connected in future.

Modelling Network Evolution

Probably social network is the fastest-growing dynamic entity in the real-world network. It is important to study the structure of network dynamics and mechanisms behind the network evolution (Grindrod et al. 2011). The structure of the social network is inherently dynamic in nature as a large number of entities are dynamically added along with their relationships. A few of the entities may also want to be disconnected from the network over time. This leads to change in several network parameters, like the number and structure of communities, node centrality, degree distribution, clustering coefficient, modularity, etc. Modelling of the dynamics can be used in generating the synthetic networks whose properties resemble a real-world social network. Scale-free network model (Holme and Kim 2002), random graph model (Aiello et al. 2000), exponential random graph model (Robins et al. 2007) and small world model (Newman and Watts 1999) are some of the widely accepted models which generate the real-world social network. A network is said to be scale-free if its degree distribution is observed to follow the power law which is already discussed in equation 1. A few nodes in the network are observed to have an unusually high degree as compared to other nodes and the degree of most of the nodes is quite less. A number of complex network systems, like citation network, social network, communication network, internet, the worldwide web are said to follow scale-free networks. Unlike scale-free network, the random model is a generative model where the degree distribution in the network is random and most importantly, it follows the small-world network. Small-world network is a network model where any two nodes in the network can be reachable by a certain number of acquaintances. Exponential random graph model (ERGM) is one kind of statistical model that tries to analyze the statistical properties of the social network (Robins et al. 2007). The statistical metric, like density, assortative, centralities capture the structural properties of the network at a specific instance of time. The ERGM model is considered to be the most suitable model for capturing the differences in statistical metrics of the network at two different instances of time. Capturing and presenting the accurate network dynamics in the large-scale network with the help of different generative network models may need a huge amount of computational time. Some big data tools, like GraphX component in Spark framework may be helpful in capturing and analyzing the huge data in order to model the real-world network (Xin et al. 2013).

Social Influence Analysis

Social influence is the behavioural change of a social entity due to relationships with other people, organizations, community, etc. Social influence analysis has attracted a lot of attention over the past few years. By analyzing the strength of influences among the people in society, a number of applications for advertising, recommendations, marketing may be developed. The strength of influence depends on many factors, like the distance between people, network size, clustering coefficient and other parameters. Since the social network is a collection of relationships between social entities, influence of one entity may affect the activity of other. The influencing factor of the entity depends on the importance of that entity in the network which can be analyzed through centrality analysis. In social influence analysis, researchers are trying to model the spread of influence, and its impact. Identifying most influential persons allows the researcher to seed the information at the proper location in order to propagate it effectively. The size of social network services like Facebook, Twitter, LinkedIn is increasing at an exponential, allowing the researchers to analyze the influence of large-scale network.

Big Data Analytics in Social Network

In the past decade, the amount of data has been created very steeply. More than 30,000 GB of data are generated every second with a great rate of acceleration. These data may be structured, unstructured or semi-structured. The sources of these data are blog posts, social network data, medical, business data, digital data, research data, scientific data, internet data, etc. The internet is the ultimate as the source of data, which is almost indecipherable. The volume of data available at present is indeed huge, but it may not be the most relevant characteristic of data. The extreme growth of data severely influences business. Business executives, scientists, medical practitioners, business executives, governments and advertising alike regularly face problems with huge data-sets in areas including finance, Internet search, business informatics and urban informatics. The main objective of today's era is simply to extract value from data by means of user behavior analytics, predictive analytics, or any other advanced data analysis method that rarely works on a specific size of data set. Traditional database systems, such as relational database management systems, are inadequate to deal with a large volume of data. These database systems face the challenges of storage, data curation, capture, search, analysis, updating, sharing, visualization, transfer, querying and information privacy (Fig. 1).

Big data mainly addresses the scalability and complexity issues that appear in a traditional dramatic fashion within a reasonable elapsed time. It mainly refers to the large and complex data sets that are not conventionally analyzed by traditional database systems. It needs a set of new techniques and technologies in order to extract useful information from datasets that are heterogeneous in nature.

Big data can also be defined in terms of 'three V's' that occur due to data management challenges (Zikopoulos et al. 2011). Three V's are presented in Fig. 2. Traditional database systems are unable to solve the issues related to these three V's. Here the 'three V's' of big data is defined as volume that ranges from MB (megabytes) to PB (petabytes) of data. The velocity defines the rate at which data are generated and delivered to the end user and variety includes data from a different variety of formats and sources (e.g., financial transactions, e-commerce, online transactions, social media interactions and weblogs, bioinformatics data, medical data, etc.).

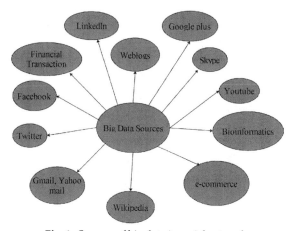

Fig. 1. Sources of big data in social network.

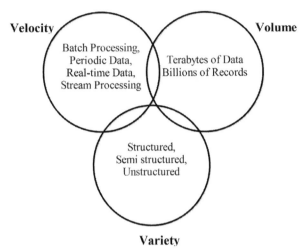

Fig. 2. 3 V's of big data analytics.

Application of Big Data Analytics

- Marketing agencies predict the future strategies by studying the information kept in the social networks like Twitter, Instagram, WhatsApp, Facebook, etc.

- Business organizations learn the reaction of their consumer regarding their products, using the information in the social media, such as product perception, preferences, etc.

- By using the historical data of the patient, analysis can be done to provide better and quick services in the healthcare centre.

- Organizations can identify the errors within a quick span of time. Real-time error detection helps organizations to react quickly in order to minimize the effects due to any operational problem. This prevents consumers from stopping use of products and save the operation from failing completely or falling behind.

- Big data analytic tools eventually save a lot of money even though expensive. They help to reduce the extra burdens of business leaders and analysts.

- Fraud detection is one of the most useful applications of big data analytics where criminals are easily identified and damage can be minimized by taking prompt measures.

- Organizations can increase their revenue by improving the quality of services by monitoring the products used by their customers.

Big Data System

Following are few of the desirable properties of the big data systems:

- *Robustness and fault tolerance*: In all circumstances, the system must behave correctly. It is essential that the system must be human-fault tolerant.

- *Low latency reads and updates*: System must achieve low latency reads and updates without compromising the robustness of the system.

- *Scalability*: Maintaining performance by adding resources to the existing system when load increases.

- *Generalization*: It must be useful in a wide range of application.

- *Extensibility*: Extra functional features can be added without changing the internal logic of the system.

- *Ad hoc queries*: Ability to mine unanticipated values from large dataset arbitrarily.

- *Minimal maintenance*: System must be maintainable (keep running smoothly) at minimum cost. One way of achieving this goal is by choosing the component with less implementation complexity.

- *Debuggability*: System must provide all essential information for debugging the system if any-thing goes wrong.

How Does Big Data Work?

With the development of modern tools and technologies, it is possible to perform different operations on larger datasets and valuable information can be extracted by analyzing the datasets. Big data technologies make it technically and economically feasible. Processing of big data involves different steps that begin from collection of raw information to implementation of actionable information.

Collection of data: Many organizations face different challenges in the beginning while dealing with big data. Such challenges are mobile devices, logs, raw data transactions, social sites, etc., but these are overcome by an efficient big data platform and developers are allowed to ingest the data from different varieties.

Storing: Preprocessing and post-processing data are stored in a secure, durable and scalable repository system. This is an essential step for any big data platform. Depending on the particular requirements, the user can opt for temporary storage of data.

Processing & Analyzing: Transformation of raw data into user consumable data through various stages, like sorting, joining, aggregating, etc. The result data sets are stored and used for further processing tasks.

Consuming & Visualizing: Finally the end user gets valuable insights from the processed data sets and according to requirements the user plan for future strategies.

Evolution of Big Data Processing

Currently basically three different styles of analysis are done on data sets to carry out multiple functions within an organization. These are as follows:

Descriptive Analysis: It helps the user to get the answer to the question: "Why it happened and what happened?"

Predictive Analysis: Probability of certain events in the feature data sets are calculated by the user, like forecasting, preventive maintenance applications, fraud detection, early alert systems, etc.

Prescriptive Analysis: User gets recommended for different events and gets the answer to the question: "What should I do if 'x' happens"?

Originally, Hadoop is the most familiar framework of big data analytics that mainly supports batch processing, where datasets of huge size are processed in bulk. However, due to time constraints new frameworks, such as Apache Spark, Amazon Kinesis, Apache Kafka, etc., have been developed to support streaming data and real-time data processing.

Traditional Approach for Data Analytics

In the traditional approach, data was stored in an RDBMS like DB2, MS SQL Server or Oracle databases and this software interacts with the sophisticated software to process the data for analysis purposes. Figure 3 presents the interaction among the major components in the traditional processing system.

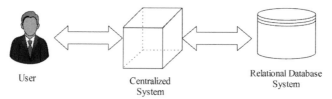

Fig. 3. Interaction in the traditional processing system.

Limitations

The traditional approach is best suited for works where the user deals with a less volume of data. While dealing with large volume of data, it was observed that the traditional database server is unable to process such data efficiently. In the traditional system, storing and retrieving volumes of data faced three major issues—as cost, speed and reliability.

Hadoop Framework

Map-Reduce programming which is found to be a most suitable approach for analyzing big data was first adopted by Doug Cutting, Mike Cafarella in their open source project named Hadoop in 2005 (Bialecki 2005). It was named after his boy's toy elephant. Hadoop is a distributed platform where data can be processed in multiple nodes in a cluster. Statistical analysis of a huge amount of data was made possible using Hadoop in a reasonable amount of time. The computational time also depends on the dependencies that exist among the data. Hadoop was written in java programming language; however, it is also compatible with python, R, and Ruby. Hadoop provides scalable and fault-tolerant platform by keeping the replica of data in multiple nodes which also increases the reliability of the programming environment. As we know, the social network data is very large and traditional programming tools are quite incapable of handling such enormous data. Hadoop is considered as a suitable programming environment for big data analytics.

Hadoop Architecture

The basic modules for Hadoop distributed framework are as follows:

- *Hadoop Map Reduce*: A programming module where data is processed in two phases—map and reduce.
- *Hadoop Common*: Common Java library files reside in this module which can be used by other modules of Hadoop. It also provides OS level abstraction and library file to start and stop programming platform.
- *Hadoop YARN*: This module contains scripts for task scheduling and resource management in the cluster of computing nodes (Vavilapalli 2013).
- *Hadoop Distributed File System (HDFS)*: It is the dedicated file system for Hadoop framework where files can be distributed across thousands of nodes. One cannot access the files of HDFS in the local file system (Shvachko et al. 2010).

Map Reduce Programming Model

Map Reduce is a programming model suitable for distributed and process the huge amount of data in several commodity nodes in the cluster simultaneously. Reliability, scalability and fault tolerant are the key features of the map reduce programming model. Each data processing in the map-reduce environment must pass through two phases, i.e., mapper and reducer. Map-reduce programming model is presented in Fig. 4.

Mapper Phase: It is the initial phase of data processing, where input data is first transformed into the set of key-value pair and passes as the argument to the mapper module. The output of the mapper phase is another set of key-value pair.

Reducer Phase: This phase takes the output of mapper as input key-value pair and reduces the size of tuple set by combining the value of each key. The output of the reducer is also a key-value pair. The reducer phase always performs after the mapper phase.

HDFS is the file system where data are distributed across multiple computing nodes. All the inputs and outputs are communicated with HDFS file systems throughout the processing steps. The framework has the responsibility for both scheduling and monitoring the task. If any task fails, it is automatically handled in a fault-tolerant manner. In the cluster of nodes, one node is considered as master and the others are treated as slaves. The framework consists of single Job Tracker for master and a Task Tracker for each slave in the cluster. The master is responsible for distributing and monitoring the task across several nodes. Job Tracker keeps track of resource consumption by all the nodes. Slave's Task Tracker executes the assigned task to the slave and provides the status of execution to the master periodically. The master is the single-point failure in Hadoop environment. If the master fails to execute, all the tasks of the slaves come to a halt.

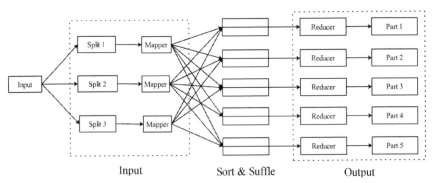

Fig. 4. Map-reduce programming model in Hadoop platform.

Advantages of Hadoop

- Hadoop is one of the most suitable tools for quickly distributing the dataset in multiple computing nodes for processing simultaneously.
- Hadoop can handle the data in a fault-tolerant and reliable manner automatically by keeping multiple blocks of the dataset in several nodes. It has a dedicated library for handling any failure at the application layer.
- The data node, i.e., one of the slave nodes can be added or deleted from the system without affecting the work in progress.
- One of the major advantages of Hadoop is that it is compatible with all the platforms. Apart from Java, it also supports a number of other languages, like Python, R, Ruby, etc.

Limitation of Hadoop

- For a better reliability and fault tolerant, multiple copies of datasets are stored in HDFS file system. It is inefficient for storage space where data size is very big.
- Hadoop is based on NoSQL database where a limited query of SQL is supported. Some open source software could make use of Hadoop framework but due to limited SQL query, they might not be that much useful.
- The execution environment is not efficient due to lack of query optimizer; size of the cluster must be large enough to handle the similar kind of database of moderate size.
- For execution in Hadoop framework, dataset must be transformed into key-value pair pattern; however, it is not possible to transform all datasets into the specified key-value pair. Therefore, it is found to be a challenging execution environment for many big data problems.
- Writing efficient program requires skill set for both data mining and distributed programming where knowledge of parallel execution environment is necessary.
- Hadoop is based on map-reduce programming model where data has to land between map and reduce phase which requires a lot of IO operation.
- Hadoop is not suitable for real-time data streaming analysis.

Challenges Faces in Hadoop Framework

- Hadoop is the distributed framework, where data are distributed across several computing nodes. The most challenging task in Hadoop is to manage the execution environment. The security model for Hadoop is by default unable to perform efficient execution. The one who wants to manage Hadoop requires to have knowledge about the security model. Encryption level at storage and network level is less tight, so huge data may be at risk in Hadoop framework.

- Hadoop is written entirely in Java language. Although Java is platform independent and widely used programming language, it is heavily vulnerable to cybercriminal activities that may lead to security breaches in Hadoop.

- Hadoop is not at all suited to small sized data. It takes more computational time as compared to traditional systems for small size data. It is not recommended for organizations where data size is not heavy.

Spark Programming Framework

Apache Spark was designed as a computing platform to be fast, general purpose and easy to use (Shoro 2015). The question is why should the user want to use Spark? As we know, Spark is closely related to map reduce in a sense that it extends on its capabilities. For batch processing, Hadoop is best, but overall, Spark is faster than others. So Spark is an open source processing engine built around speed, ease of use and analytics. If we have a large amount of data that requires low latency processing that a typical map reduces program cannot provide, then Spark is the way to go. Generally, computations are of two types—one is I/O intensive and another is CPU intensive. Spark performance seems to be much better for I/O intensive computation. In Spark, data is loaded in a distributed fashion in memory and is free from data locality concept. Spark is completely in-memory processing; hence it is faster than Hadoop.

Like map reduces, Spark provides parallel distributed processing, fault tolerance on commodity hardware, scalability, etc. Spark tool is time- and cost-efficient due to the concept called cached in memory distributed computing, high-level APIs, low latency and stack of high-level tools. Mainly two groups of people use Spark—one is data scientist and another is engineer.

Data scientists analyze and model the data to obtain information. They use some techniques to transform the data into some suitable form so that they can easily analyze. They use Spark to get the results immediately. Once the hidden information is extracted from the data, later the persons who work on this hidden information are called engineers. Engineers use Spark's programming APIs to develop business applications while hiding the complexities of distributed system programming and fault tolerance across the clusters. Engineers can also use Spark for monitoring and inspecting applications.

Component of Spark Framework

Spark core is a general purpose system that provides scheduling, distributing and monitoring of the applications in a cluster. On top of the core, components are designed to interoperate closely, so that users can combine them as a library in a software project. Top-level components will inherit the improvements made at the lower layers, e.g., by optimizing the Spark core speed of SQL, machine learning, the streaming and graph processing libraries increase. Spark core can even run with its own built-in scheduler. It also can run over a different cluster manager, like apache mesos and Hadoop YARN. Spark components reside above the Spark core are shown in Fig. 5.

Spark SQL: It works with unified data access which means the user can have any kind of data. Since it is compatible with Hive, the user can run hive queries on top of the Spark SQL. As a result, time is saved on migrating hive queries to Spark queries and no conversion is required. Queries are embedded in Python, Scala and Java. It gives a lot of tight integration with SQL while providing standard connectivity for all popular tools, like ODBC, JDBC.

Spark Streaming: Streaming task means immediately processing data and generating the result at runtime. It is a real-time language integrated APIs to do stream processing by using Java and Scala. This stream processing framework is fault tolerant because Spark has got an excellent recurring mechanism. The user can also combine stream processing with another component, like MLib, GraphX, etc. (Meng et al. 2016, Gonzalez et al. 2014). User can run streaming with both batch and interactive mode. Streaming

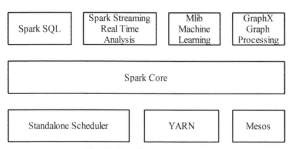

Fig. 5. Spark components.

also integrates with lots of existing standards like HDFS, FLUME, KAFKA (Hoffman 2013, Garg 2013). Kafka is distributed message queue which is used for data processing in streaming work. It was written by Twitter. Flume framework works on top of Hadoop. Hadoop performs poor in streaming data and does not have provision for processing streaming data.

Spark MLib: It is a collection of machine learning library functions and runs 100 times faster than Hadoop because of Spark storing data in memory. It has a lot of machine learning algorithm like *k*-means clustering, linear regression, logistic regression, etc. Spark optimizes the algorithms to run top of Spark platform. One main feature of this component is that it can deploy existing Hadoop clusters. In Hadoop, there is no provision for built-in machine learning algorithm. The user has to use Mahout Platform to build an application in machine learning (Lyubimov 2016).

RDD: RDD stands for Resilient Distributed Data which is a basic abstraction unit of Spark. Spark provides APIs to interact directly with RDD, which has a higher partition tolerance as compared to HDFS. In HDFS the user gets partition by doing replication. But in RDD the replication factor is lot less as compared to HDFS, so it requires less space. RDD is immutable and Spark stores the data in RDD.

GraphX: Spark provides an optimized framework for graph processing. An optimized engine supports the general execution of graphs called DAGs. A lot of built-in algorithms are present for graph processing. GraphX is a unique feature in Spark (Gonzalez et al. 2014).

Functional features of Spark

- It provides powerful caching and disk persistence capabilities.
- Faster batch processing as compared to Hadoop.
- Real-time stream processing.
- Faster decision making, because data resides in memory. Hence the user can definitely go for iterative algorithms. This is a severe limitation in Hadoop.
- It supports iterative algorithms. In Hadoop, sharing of data is not possible which leads to non-iterative algorithms.
- Spark is interactive data analysis, whereas Hadoop is non-interactive.

Non-functional Features of Spark

- A fully apache hive compatible data warehousing system that can run 100x faster than Hive.

- Spark framework is polyglot, or it can be programmed in several programming languages (currently in Java, Python and Scala).
- Spark is not really dependent on Hadoop for its implementation as it has got its own manager.

What is Streaming Data?

Because of the dynamic nature of social network, data is generated continuously at an exponential rate. Real-time data that stream in the era of big data can be considered as streaming data. They can include a variety of data in different sizes. This homogenous nature of data makes the analysis more complex. The major source for streaming data comes from the log files that are generated from e-commerce website, online gaming activity, message exchanges by the user in the online social network, weather forecasting, information generated from health care systems, etc. Streaming data must be analyzed on a record-by-record basis over a time-frame. A wide range of analyses like regression analysis, correlation, aggregation, sampling, filtering can be done on real-time streaming data. Visualization of information gathered from these analyses can be utilized to improve the marketing business by responding to the upcoming situation effectively. For example, companies can analyze the sentiment data over their e-commerce product for building a better recommendation-engine for their product.

Benefits of Streaming Data

Streaming data processing is helpful in many situations where new, dynamic information is produced consistently. It applies to a majority of the business portions and huge information utilize cases. Organizations, for the most part, start with basic applications, like gathering framework logs and simple handling, like moving min-max calculations. Hence, these applications develop to more complex real-time processing. At first, applications may handle information streams to deliver basic reports and perform basic activities accordingly, for example, discharging cautions when key measures surpass a particular threshold. In the end, these applications perform more refined types of information examination, such as machine learning algorithm applications and gather further bits of knowledge from information. Further, stream, complex and algorithms involving event processing, such as decaying time windows to locate the latest prominent motion pictures, are connected, additionally enhancing the knowledge.

Examples of Streaming Data

Sensors in vehicles used for transportation, modern hardware and machines used for farming send information to a streaming application. The application screens execution performance and identifies any potential errors ahead of time, and puts in an extra part order request that prevents equipment downtime.

- Changes in the share trading system can be tracked progressively by a financial institution, which naturally, rebalances portfolios in view of stock value developments.

- A subset of information from customers' cell phones is tracked by the real-estate websites and continuous property suggestions of properties to visit in light of their geolocation is made.

- A company using solar energy to replenish needs to keep up power throughput for its clients or pay penalty. It thus implements an application based on streaming data analysis which scrutinizes all the panels in the field, scheduling the service in real-time, which lessens the duration of throughput from each and every panel along with its associated penalty payouts.

- Humongous numbers of clickstream records are streamed by media publishers taken from their online contents, thus aggregating and enriching the data with certain location-related information about its users. They also optimise the placement of the content on its site, which delivers a better and relevant experience to its users.

- Streaming data that involves the interaction between a player and a game can be collected by an online gaming company. This collected data is fed into their platform which analyses real-time data and offers new experiences to the players who play it which gives them a better liking for the game.

Comparison between Batch Processing and Stream Processing

Before diving into the arena of streaming data, we can have a look at the differences between two major areas of processing techniques, namely, batch processing and stream processing. Batch processing can be utilized to figure random inquiries over various information. It computes results which are computed from all the already encompassed data, thus enabling a deep and close analysis of big datasets. Amazon EMR, which is a map-reduce-based system is an example of a platform which supports batch jobs. To the contrast, ingesting a sequence of data is something required by stream processing. Incremental updating metrics, summary statistics and reports in response to each arriving data record are also sometimes a necessity for stream processing which makes it more efficient for real-time processing and other responsive functions.

Table 1. Comparison between batch processing and stream processing.

	Batch Processing	Stream Processing
Data scope	Uses all or most of the data in the dataset for querying and processing.	Uses data within a very small time window or on just the most recent data record for processing and querying.
Data size	Huge amount of data, in the form of batches of data, is used.	Single records or very small batches called 'micro-batches' of records are used.
Performance	Offers more latency which last in minutes to hours.	Provides less latency which is in the order of few seconds or milliseconds.
Analyses	Analyses of these data are quite complex.	Analysis is done through simple response functions, aggregates, and rolling metrics.

Challenges in Analyzing Streaming Data

There are two layers for streaming data processing—a processing layer and a storage layer. There lies a need to support ordering of records and strong consistency to have a faster, less costly and reads and writes of large streams of data which are repayable by the storage layer. The second layer, that is the processing layer, is responsible for consumption of data from the storage layer, executing various actions on that data and finally, notifying the storage layer to remove the data that is no longer required. Along with all the already available features, we also need to have a plan which ensures fault tolerance, data durability and scalability in both streaming data processing layers namely, the storage and processing layers. There have been many platforms that have been developed as a result of this, which provide the platform and the framework required to build streaming data applications. Some of these are Apache Storm, Apache Flume, Apache Spark Streaming and Apache Kafka.

Conclusion

Social network analysis is a typical example of an idea that can be applied in many fields. With mathematical graph theory as its basis, it has become a multidisciplinary approach with applications in sociology, the information sciences, computer sciences, geography, etc. In this chapter, different research direction in social network analysis has been explained with relevant application. As the data generated from social network is increasing at an exponential rate, the traditional system fails to process and analyze such huge heterogeneous and complex data. Hadoop or Spark may be considered as an alternative approach for analyzing such data as they have an individual speciality in analyzing the social structure and social media data. Hadoop is basically used to analyze batch mode social data, whereas Spark may be considered as suitable for real-time streaming data.

References

Aggarwal, Charu C. (2011). An Introduction to Social Network Data Analytics. In Social Network Data Analytics. Springer, Boston, MA, 1–15.

Aiello, William, Fan Chung and Linyuan, Lu. (2000). A random graph model for massive graphs. Proceedings of the Thirty-second Annual ACM Symposium on Theory of Computing 171–180.

Barabsi, Albertszl and Rka Albert. (1999). Emergence of scaling in random networks. Science (American Association for the Advancement of Science) 286(5439): 509–512.

Barthelemy, Marc. (2004). Betweenness centrality in large complex networks. The European Physical Journal B-Condensed Matter and Complex Systems 38(2): 163–168.

Bialecki, Andrzej. (2005). Hadoop: a framework for running applications on large clusters built of commodity hardware. http://lucene. apache. org/hadoop.

Borgatti, Stephen, P. and Martin G. Everett. (2006). A graph-theoretic perspective on centrality. Social Networks 28(4): 466–484.

Brandes, Ulrik. (2001). A faster algorithm for betweenness centrality. Journal of Mathematical Sociology 25(2): 163–177.

Day, William H.E. and Herbert Edelsbrunner. (1984). Efficient algorithms for agglomerative hierarchical clustering methods. Journal of Classification 1(1): 7–24.

Fortunato, Santo. (2010). Community detection in graphs. Physics Reports 486(3): 75–174.

Freeman, Linton C. (1978). Centrality in social networks conceptual clarification. Social Networks 1(3): 215–239.

Garg, Nishant. (2013). Apache Kafka. Packt Publishing Ltd. Birmingham, United Kingdom.

Gonzalez, Joseph E., Reynold S. Xin, Ankur Dave, Daniel Crankshaw, Michael J. Franklin and Ion Stoica. (2014). GraphX: Graph processing in a distributed dataflow framework. OSDI 599–613.

Grindrod, Peter, Mark C. Parsons, Desmond J. Higham and Ernesto Estrada. (2011). Communicability across evolving networks. Physical Review E 83(4): 046120.

Gulati, Ranjay. (1995). Social structure and alliance formation patterns: A longitudinal analysis. Administrative Science Quarterly 619–652.

He, Wu, Shenghua Zha and Ling Li. (2013). Social media competitive analysis and text mining: A case study in the pizza industry. International Journal of Information Management 33(3): 464–472.

Hey, Tony, Stewart Tansley, Kristin M. Tolle and others. 2009. The Fourth Paradigm: Data-intensive Scientific Discovery. Vol. 1. Microsoft Research Redmond, WA.

Hoffman, Steve. (2013). Apache Flume: Distributed Log Collection for Hadoop. Packt Publishing Ltd.

Holme, Petter and Beom Jun Kim. (2002). Growing scale-free networks with tunable clustering. Physical Review E 65(2): 026107.

Kleinfeld, Judith S. (2002). The small world problem. Society 39(2): 61–66.

Knoke, David and Song Yang. (2008). Social Network Analysis. Vol. 154. Sage.

Liben-Nowell, David and Jon Kleinberg. (2007). The link-prediction problem for social networks. Journal of the Association for Information Science and Technology 58(7): 1019–1031.

Lyubimov, Dmitriy and Palumbo, Andrew. (2016). Apache Mahout: Beyond MapReduce. CreateSpace Independent Publishing Platform.

Meng, Xiangrui, Joseph Bradley, Burak Yavuz, Evan Sparks, Shivaram Venkataraman, Davies Liu, Jeremy Freeman et al. 2016. Mllib: Machine learning in apache spark. The Journal of Machine Learning Research 17(1): 1235–1241.

Newman, Mark E.J. and Duncan J. Watts. (1999). Scaling and percolation in the small-world network model. Physical Review E 60(6): 7332.

Newman, Mark E.J. and Michelle Girvan. (2004). Finding and evaluating community structure in networks. Physical Review E 69(2): 026113.

Newman, Mark E.J. (2005). A measure of betweenness centrality based on random walks. Social Networks 27(1): 39–54.

Newman, Mark E.J. (2008). The mathematics of networks. The New Palgrave Encyclopedia of Economics 2: 1–12.

Okamoto, Kazuya, Wei Chen and Xiang-Yang Li. (2008). Ranking of closeness centrality for large-scale social networks. Lecture Notes in Computer Science 5059: 186–195.

Pang, Bo, Lillian Lee and others. 2008. Opinion mining and sentiment analysis. Foundations and Trends{\textregistered} in Information Retrieval (Now Publishers, Inc.) 2(12): 1–135.

Rahwan, Talal, Michalak, Tomasz P. and Wooldridge, Michael. (2016). Closeness centrality for networks with overlapping community structure. Thirtieth AAAI Conference on Artificial Intelligence. Arizona, USA.: AAAI Publications 622–629.

Reichardt, Jiorg and Stefan Bornholdt. (2006). Statistical mechanics of community detection. Physical Review E 74(1): 016110.

Robins, Garry, Pip Pattison, Yuval Kalish and Dean Lusher. (2007). An introduction to exponential random graph (p*) models for social networks. Social Networks 29(2): 173–191.

Ruhnau, Britta. (2000). Eigenvector-centrality, a node-centrality? Social Networks 22(4): 357–365.

Shoro, Abdul Ghaffar and Tariq Rahim Soomro. (2015). Big data analysis: Apache spark perspective. Global Journal of Computer Science and Technology 15(1).

Shu, Wesley and Yu-Hao Chuang. (2011). The perceived benefits of six-degree-separation social networks. Internet Research (Emerald Group Publishing Limited) 21(1): 26–45.

Shvachko, Konstantin, Hairong Kuang, Sanjay Radia and Robert Chansler. (2010). The hadoop distributed file system. Mass Storage Systems and Technologies (MSST), 2010 IEEE, 26th Symposium on. 1–10.

Tang, Lei and Huan Liu. (2010). Community detection and mining in social media. Synthesis Lectures on Data Mining and Knowledge Discovery (Morgan & Claypool Publishers) 2(1): 1–137.

Vavilapalli, Vinod Kumar, Arun C. Murthy, Chris Douglas, Sharad Agarwal, Mahadev Konar, Robert Evans, Thomas Graves et al. (2013). Apache hadoop yarn: Yet another resource negotiator. Proceedings of the 4th Annual Symposium on Cloud Computing. Santa Clara, California: ACM, 5.

Watts, Duncan J. and Steven H. Strogatz. (1998). Collective dynamics of small-world networks. Nature (Nature Publishing Group) 393(6684): 440–442.

White, Tom. (2012). Hadoop: The Definitive Guide. O'Reilly Media, Inc.

Xin, Reynold S., Joseph E. Gonzalez, Michael J. Franklin and Ion Stoica. (2013). Graphx: A resilient distributed graph system on spark. First International Workshop on Graph Data Management Experiences and Systems. 2.

Zhang, Jing, Jie Tang and Juanzi Li. (2007). Expert finding in a social network. International Conference on Database Systems for Advanced Applications 1066–1069.

Zikopoulos, Paul, Chris Eaton and others. (2011). Understanding Big Data: Analytics for Enterprise Class Hadoop and Streaming Data. McGraw-Hill Osborne Media.

CHAPTER **2**

An Analysis of AI-based Supervised Classifiers for Intrusion Detection in Big Data

Gulshan Kumar[1], and *Kutub Thakur[2]*

Introduction

Our dependence on Internet-based applications is increasing day by day for fulfilling daily requirements. The Internet and online operations have become a significant part of our daily life as they constitute an important component of today's business scenario (Shon and Moon 2007). This Internet connectivity and increasing dependence of business applications enables malicious users to misuse resources and mount a variety of attacks. Since the last decade, there is certainly the technological push at the infrastructure level. Nowadays, Internet-access speed of 1–10 Gbps is usual. This technological improvement has paved the way to new challenges. Firstly, the amount of Internet traffic, as well as the line speed, continues to grow. As depicted in Fig. 1, there is an almost exponential growth in the last decade, with occasional peaks up to Gbps (Internet2 2010). Such large amounts of data from multiple information sources need

---

[1] Department of Computer Applications, Shaheed Bhagat Singh State Technical Campus, Ferozepur, Punjab, India.
[2] Department of Professional Security Studies, Cyber Security, New Jersey City University, United States; Email: kthakur@njcu.edu
* Corresponding author: gulshanahuja@gmail.com

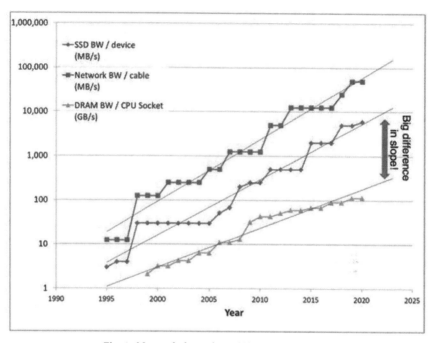

Fig. 1. Network throughput (Gbps) for network.

to be managed, monitored and analyzed. This requires development of new strategies, like big data analytics to cope with an average load of multiple Gbps. Secondly, the number of attacks also continues to grow. The reason behind this is in itself very simple—attacks are getting economically more and more profitable (Sperotto 2010). The continuously growing Internet attacks and large amount of Internet traffic consisting of data of multiple source of information pose a severe challenge to develop a flexible, adaptive security-oriented technique. DARPA divides the different attacks and intrusions into four different classes viz: (1) Probe; (2) DoS (Denial of Service); (3) U2R (User to Root); (4) R2L (Remote to Local) (MIT 2001). Various security-oriented technologies have been developed in the last decade to prevent and detect the ever-growing attacks in the big data. Nowadays, Intrusion Detection System (IDS) is a key component as security mechanism for detection of cyber attacks. An intrusion is defined as any set of actions that attempt to compromise the integrity, confidentiality or availability of information resources (Hernandez-Pereira et al. 2009). An intrusion detection system (IDS) is an effective security technology, which can detect such malicious actions (Halme 1995).

It monitors and analyses the audit trails and network traffic on the systems and networks respectively to provide security services. The main objective of IDS is to detect all possible intrusions in an efficient manner and notify the network security administrator. One major challenge to IDS is to analyze a large amount of audit patterns and network traffic. The different features from network connections are recorded in the form of network connection records. These features involve irrelevant and redundant features. Processing and analysis of irrelevant and redundant features leads to (1) undesirable delay in classification task, which in turn loses the real time capability of IDS; (2) increase in computation overhead in terms of memory and time; and (3) deterioration in the classification accuracy.

To cope with this problem, various feature reduction techniques are applied to remove irrelevant and redundant features. In this work, we utilized our earlier proposed information theoretic approach for feature selection to reduce the number of features (Kumar and Kumar 2012). The reduced dataset containing relevant features has been analyzed for intrusion detection. Various techniques from different disciplines were applied in intrusion detection to classify a wide variety of ever-growing cyber intrusions. Due to advantages of Artificial Intelligence (AI)-based intrusion detection techniques over other conventional techniques has made it focus on current research in security domain (Ponce 2004). Witten and Frank (2005) divided various AI-based supervised classifiers into different categories, viz: (1) Rule based; (2) Tree based; (3) Functions; (4) Lazy; (5) Bayes; and (6) Meta (Witten and Frank 2005). In this work, we compared the performance of various supervised classifiers in different categories based on defined performance metrics, using benchmarked KDD 1999 dataset (KDD99 1999).

Evaluation of classifiers on a variety of metrics is very significant because different classifiers are designed to optimize different criteria. For example, SVM are designed to minimize the structural risk minimization and hence optimize the accuracy, whereas neural network are designed to minimize empirical risk minimization and hence optimize root mean squared error (RMSE). It is common that one classifier may show optimal performance on one set of metrics and suboptimal on another set of metrics. The identified standard performance metrics involve F-measure (FM), classification rate (CR), false positive rate (FPR), cost per example (CPE), precision (PR), root mean square error (RMSE), area under ROC curve (ROC) and detection rate (DR). Based upon these popularly used metrics, we propose a new single generalized metric, called GPM, which may be used to measure the overall performance when no specific metric is known. We identify the best classifier for each attack class in the respective classifier category. Further, we compared best classifiers in the respective category to identify overall the best classifier in different class attacks.

The major contributions of this work are:

1) Proposal of a generalized performance metric based upon identified metrics to measure overall performance of the classifiers.

2) Employment of information theoretic feature-selection technique to reduce the features of intrusion detection KDD 1999 dataset.

3) Generation of the trained model of various classification techniques, using reduced training KDD 1999 dataset.

4) Empirical comparison of various classifiers in different classifier categories to identify best classifier in the respective category for different attack classes, using reduced test KDD 1999 dataset in terms of defined performance metrics.

5) Empirical comparison of various best classifiers of different classifier categories to identify overall best classification technique for different attack classes, using reduced test KDD 1999 dataset in terms of defined performance metrics.

To that end, we analyze the performance of supervised classifiers and evaluate their performance for intrusion detection using subsets of benchmarked KDD cup 1999 dataset as training and test dataset.

To the best of our knowledge, this empirical work is the most comprehensive analysis in terms of the number of classifiers considered and a number of metrics used for comparison with big data for network-based intrusion detection.

The rest of the paper is organized as follows: We analyzed related literature in the succeeding section. Thereafter, we describe the experimental setup and methodology for conducting experiments. This includes preparation of the evaluation dataset, employment of feature reduction, definition of performance metrics and AI-based classifier categories. The next section gives the empirical study of various classifiers in terms of five defined metrics. It provides the results of best classifiers and their analysis in terms of defined metrics. Finally, the concluding remarks and future research guidelines are highlighted in the last section.

Literature Analysis

Many researchers utilized AI-based techniques for efficient intrusion detection. These techniques include Decision Trees, Naive Bayes, Bayesian Network, Neural Network, Support Vector Machine, Nearest Neighbor technique, Random Forest and many more. Given below is the comparative work done by different researchers.

Sabhnani and Serpen (2003) conducted a number of experiments with hybrid systems that contained different approaches for attack classification at the University of Toledo (Sabhnani and Serpen 2003). They evaluated a set of pattern recognition and machine-learning algorithms. KDD Cups data was chosen for the experiments. They highlighted that most researchers employ a single algorithm to detect multiple attack categories with a dismal performance in some cases. So, they proposed to use a specific detection algorithm that is associated with an attack category for which it is most promising. They experimented with nine different classifiers and suggested different classifiers for each attack category, based on detection rate and false alarm rate as a performance metric. Finally, they proposed a multi-classifier approach for intrusion detection.

Pan et al. (2003) compared a hybrid neural network and Decision Tree model. They sampled different attacks from each class and concluded that the hybrid model is a better approach (Pan et al. 2003).

Mukkamalla and Andrew (2004) analyzed SVM, ANN, LGP (Linear Genetic Programs) and MARS (Multivariate Adaptive Regression Splines) for classification of KDD data into five classes Mukkamala et al. (2005). They used detection accuracy, attack severity, training and testing time (scalability) as evaluation metrics and used DARPA intrusion detection dataset. They proved that none of the single classifiers perform well in all attack classes. LGP outperforms MARS, SVM and ANN in terms of detection accuracy at the expense of time. MARS is superior to SVM in respect to classifying the most important classes (U2R and R2L) in terms of the attack severity. SVM outperform ANN in important respects of scalability (SVM can train a larger number of patterns, while ANN takes a long time to train or fail to converge at all when the number of patterns gets large), training time, running time (SVM runs an order of magnitude faster) and prediction accuracy.

Amor et al. (2004) compared Naive Bayes and Decision Tree-based techniques to detect intrusions. The authors used KDD99 as benchmark dataset (Amor et al. 2004). The authors conducted three types of experiments—first type detects all attacks in dataset; second type detects four different types of classes of attacks (i.e., Probe, DoS, U2R, R2L) and the last type of experiment detects only two types of attacks (i.e., normal or attack). They concluded that detecting the number of classes of attacks does not make a considerable effect on the detection process, using the above-mentioned classification techniques.

Zhang and Zulkernine (2005) applied Random forest approach to detect the intrusion in network with improved performance (Zhang and Zulkernine 2005). They balanced the dataset by changing the sampling size of minority and majority classes. They concluded that this approach can improve the detection rate of minority classes.

Xu (2006) utilized the machine-learning approach to propose a framework for adaptive intrusion detection (Xu 2006). Multi-class Support

Vector Machine (SVM) is used to classify intrusions and the performance of SVMs is evaluated on the KDD99 dataset. Results reported are: 76.7 per cent, 81.2 per cent, 21.4 per cent and 11.2 per cent detection rate for DoS, Probe, U2R and R2L respectively by keeping False Positive (FP) at relatively low levels of the average 0.6 per cent in the four categories. But the approach was evaluated using a small set of data (10,000 randomly sampled records) from huge original dataset (5 million audit records). Therefore, it is difficult to prove the effectiveness of the approach.

Gharibian and Ghorbani (2007) compared the supervised probabilistic and predictive machine-learning techniques for intrusion detection (Gharibian and Ghorbani 2007). Two probabilistic techniques—Naive Bayes and Gaussian and two predictive techniques, Decision Tree and Random Forests, are employed. They utilized KDD dataset for evaluation and detection accuracy was used as the main performance metric. It was concluded that probabilistic techniques are better than predictive techniques in the case of DoS attacks.

Panda and Patra (2008) advocated that there is no single best algorithm to outperform others in all situations (Panda and Patra 2008). They demonstrated the performance of three well-known data-mining classifier algorithms, namely ID3, J48 and Naïve Bayes evaluated based on the ten-fold cross validation test of KDD dataset, using WEKA (Weka 2011). The performance metrics include time to build the model, error rate, confusion matrix, ROC, recall rate vs. precision rate and kappa statistics.

Singh (2009) performed experiments on a subset of the KDD dataset by random sampling of 49,402 audit records for the training phase and 12,350 records for the testing phase (Singh 2009). The author reported an average TP of 99.6 per cent and FP of 0.1 per cent, but, did not provide any information about classification or detection rate for different attack classes.

Literature analysis done in the aforementioned text showed the following findings:

- Most of the researchers used subset of KDD99 cup dataset for comparison, but the preprocessing done by all the researchers in their work is not the same. Hence the work of different authors cannot be critically evaluated on the same platform.

- The classifiers used by most of the researchers for comparison are not comprehensive and sufficient, covering all categories of classifiers.

- Most of the researchers utilized all the features of KDD 99 cup dataset leading to the processing of irrelevant and redundant features. Thus the reporting results are not comprehensive for all classifiers.

- A comprehensive analysis of different classification techniques is not done, using the same set of defined performance metrics.

The main aim of this work is to (1) reduce the features of dataset using information theoretic filter feature selection approach; (2) propose a generalized metric for measuring the performance of classifier; (3) compare the performance of a sufficient number of AI-based supervised classifiers of different categories for intrusion detection, using the same preprocessed dataset in terms of five defined performance metrics; (4) identify best classifier for intrusion detection categorywise and on the whole for different attack classes.

Experimental Setup and Methodology

This section describes the evaluation dataset, preprocessing strategy, selection of training and testing dataset, formation of reduced training and testing dataset by employing of feature-selection approach, various classifier categories and other experimental setups.

Methodology

We performed experiments on Intel PIII 239 MHz with 1 GB RAM with Windows XP operating system. We performed 5-class classification of dataset using well-known open source publicly available machine-learning tool called Weka (Weka 2011) to classify KDD cup 1999 dataset. We conducted a set of experiments using default parameters of Weka-implemented classifiers. The stages of experiment and their interaction is described as follows and depicted in Fig. 2.

i. *Preprocessing stage*: In this stage, conversion of symbolic features to numeric features and normalization of features is performed for Training and Test KDD dataset as described in (Kumar et al. 2010a).

ii. *Feature reduction stage*: In this stage, an information theoretic feature selection approach (Kumar and Kumar 2012) is applied to normalized training and test dataset for generating a reduced feature set; hence, reduced training and test dataset with reduced features.

iii. *Classification stage*: Classification stage involves two phases—namely, training phase and testing phase.

   a. *Training Phase*: Here, the classifier is learnt, using reduced training dataset. The output of this phase is a trained model which optimized using 10 cross validation.

   b. *Testing Phase*: Here, the trained model is given an input of test dataset to predict the class label.

iv. *Performance metrics computation*: After the testing phase, performance metric computation stage computes the defined performance metrics. We divided the performance metrics into three classes: threshold,

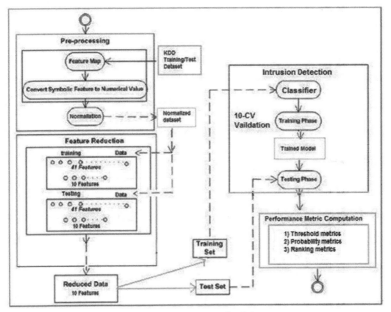

Fig. 2. Experiment methodology.

ranking and probability metrics (Caruana and Niculescu-Mizil 2004). Threshold metrics include classification rate (CR), F-measure (FM) and cost per example (CPE). It is not important how close a prediction is to a threshold, but, only if it is above or below threshold. The value of threshold metrics lies in (0, 1). Ranking metrics include false positive rate (FPR), detection rate (DR), precision (PR) and area under ROC curve (ROC). The value of ranking metrics lies in (0, 1). These metrics depend on the ordering of the cases; not the actual predicted values. As long as the ordering is preserved, it makes no difference. These metrics measure how well the attack instances are ordered before normal instances and can be viewed as a summary of the model performance across all possible thresholds. Probability metrics include root mean square error (RMSE). The value of RMSE lies between 0 and 1. The metric is minimized when the predicted value for each attack class coincides with the true conditional probability of that class being normal class.

These metrics are computed from confusion matrix. The matrix gives the values of True Positives (TP), True Negatives (TN), False Positives (FP) and False Negatives (FN). The details of these metrics can be further elaborated in Appendix A.

We propose a generalized performance metric (GPM) to measure the overall performance of the classifier as follows:

$$GPM = \frac{\alpha1*CR + \alpha2*(1-CPE) + \alpha3*FM + \alpha4*(1-RMSE) + \alpha5*DR + \alpha6*(1-FPR) + \alpha7*ROC + \alpha8*PR}{\sum_{i=1}^{8}\alpha_i}$$

where α_i is the average Spearman rank correlation of ith metric with the other metrics. To compute the values of α_i, we computed the Spearman rank correlation between different metrics, using PASW statistics18 (IBM SPSS 2011). The values are depicted in Table 1. The last column gives an average correlation of the metrics with other metrics. Based upon these values, we compute GPM, whose value lies between 0 and 1. Higher the value of GPM, better the performance of classifier.

Table 1. Spearman rank correlation of different metrics.

	CR	1-CPE	FM	1-RMSE	DR	1-FPR	ROC	PR	GPM	Mean
CR	1.000	0.851	0.979	0.523	0.998	0.493	0.325	0.516	0.858	0.669
1-CPE	0.851	1.000	0.909	0.523	0.840	0.647	0.484	0.663	0.935	0.702
FM	0.979	0.909	1.000	0.481	0.971	0.556	0.372	0.565	0.582	0.690
1-RMSE	0.523	0.523	0.481	1.000	0.494	0.578	0.568	0.607	0.843	0.539
DR	0.998	0.840	0.971	0.494	1.000	0.466	0.312	0.491	0.778	0.653
1-FPR	0.493	0.647	0.556	0.578	0.466	1.000	0.700	0.955	0.625	0.628
Avg ROC	0.325	0.484	0.372	0.568	0.312	0.700	1.000	0.620	0.898	0.483
PR	0.516	0.663	0.565	0.607	0.491	0.955	0.620	1.000	0.772	0.631
WT GPM	0.858	0.935	0.582	0.843	0.778	0.625	0.898	0.772	1.000	0.786

Intrusion Detection Dataset

We used well-known intrusion detection benchmarked KDD cup99 dataset to compare and analyze different classification techniques. This dataset contains data of connection records consisting of network header data, network flows, logs, and system events. Network flows, logs and system events, etc., generate big data. Big data analytics using AI-based techniques can correlate multiple information sources into a coherent view, identify intrusive activities and finally achieve effective and efficient intrusion detection.

Although KDD99 data-set might have been criticized for its potential problems (Tavallaee et al. 2009, McHugh 2000), but many researchers give the priority to KDD dataset over other publicly available dataset as benchmark dataset for evaluation of IDS (Tsai et al. 2009). KDD dataset utilizes TCP/IP level information and embedded with domain-specific heuristics, to detect intrusions at the network level. It contains four major classes of attacks: Probe, Denial of Service (DoS), User-to-Root (U2R) and Remote-to-Local (R2L) attacks. There are 4,898,430 labeled and unlabeled and 3,11,029 unlabeled connection records in the dataset. The labeled connection records consist of 41 features and 01 attack type. The number of connection records in the training set and the test dataset is very large and non-uniformly distributed for different attack classes. In order to perform unbiased normalized training and testing of classifier model, we randomly choose the connection records as suggested by Kumar et al. (2010). There are 66,961 (including normal) connection records selected from an entire labeled KDD dataset for training of classifiers. There are 40,603 connection records in the test dataset. KDD-dataset contains symbolic as well as continuous features. The dataset is pre-processed to make it compatible before it is used for training and testing the classifiers. The pre-processing steps involve mapping of symbolic value features to numeric values and normalizing of feature values as suggested by Kumar et al. (2010a).

Feature Reduction

The KDD dataset contains 41 features and 01 class features. All the features are equally significant to predict the class label. There exist some irrelevant and redundant features, processing of which leads to many problems, including (1) undesirable delay in classification task which in turn loses the real time capability of IDS; (2) increase computation overhead in terms of memory and time; and (3) deterioration in the classification and prediction accuracy. To solve this problem, we employed an information theoretic approach for feature selection suggested by Kumar and Kumar (2012). This feature selection approach is a filter approach and independent of any classification technique. Thus, KDD dataset is reduced by using this approach. The reduced KDD dataset is not dependent upon any classification technique. Here, we utilized the mutual information to compute the relevance of features to predict the class labels. The reduced KDD dataset contains 10 features of 66,961 instances in training dataset and 10 features of 40,603 instances in test dataset. Details of the number of instances in training and test dataset for various attack classes are as described in Table 2.

Table 2. Statistics of subsets of KDD cup 1999 dataset as Training and Test dataset.

	Attack Class	Number of Instances
Training dataset	Normal	10,000
	Probe	32,316
	DoS	23,467
	U2R	52
	R2L	1,126
Total		66,961
Test dataset	Normal	5,000
	Probe	4,166
	DoS	17,761
	U2R	228
	R2L	13,448
Total		40,603

AI-based Supervised Classifier Categories

Witten and Frank (2005) divided various supervised classifiers into different categories viz: (1) Rule based; (2) Tree based; (3) Functions; (4) Lazy; and (5) Bayes (Witten and Frank 2005). We compared and analyzed the following classifiers in their respective categories:

1. Tree-based classifiers: Under this category, we utilized classifiers: (1) Random Forest; (2) Random Tree; (3) Naïve Bayesian Tree (NBTree); (4) J48; and (5) Classification and Regression Tree (simple CART).

2. Rule-based classifiers: Under this category, we utilized classifiers: (1) JRIP; and (2) Decision Table.

3. Bayes classifiers: Under this category, we utilized classifiers: (1) Bayesnet; and (2) Naïve Bayes.

4. Functions-based classifiers: Under this category, we utilized classifiers: (1) MLP; (2) Sequential Minimal Optimization (SMO); (3) RBF network; and (4) LibSVM classifier.

5. Lazy classifiers: (1) Instance-based Learner (IB1); (2) IBK; and (3) KStar.

6. Meta classifiers: (1) Bagging; (2) Boosting; (3) Stacking; and (4) Random Subspace.

The details of these classifiers can be further investigated in (Witten and Frank 2005, Kumar et al. 2010b).

Empirical Study

This section presents the comparative performance evaluation of various classifier categories. Classifiers are compared in terms of eight identified standard metrics and the new proposed metric GPM.

Tree-based Classifiers

We compared: (1) Random Forest; (2) Random Tree; (3) NB Tree; (4) J48; (5) Simple CART classifiers. The comparative results are shown in Table 3. It is observed that J48 outperformed the other classifiers in terms of threshold metrics. In terms of probability metrics, random forest classifier outperformed the others and J48 is in second position. For ordering metrics like DR, FPR and PR, J48 show better performance than other classifiers. But in terms of ROC, random forest classifier outperformed J48 by putting it in second position. Last row indicates that on the whole, J48 classifier provides superior performance than other classifiers in the category of tree-based classifiers.

Table 3. Comparative results of tree based classifiers.

Tree-based Techniques/Metrics	Random Forest	Random Tree	NB Tree	J48	Simple CART
CR	0.5424	0.5987	0.5103	0.6485	0.6351
CPE	0.929	0.802	0.883	0.716	0.7184
RMSE	0.3471	0.4005	0.4227	0.373	0.3786
Avg DR	0.542	0.599	0.51	0.648	0.635
Avg FPR	0.075	0.064	0.073	0.049	0.088
Avg ROC	0.829	0.768	0.798	0.792	0.796
Avg FM	0.537	0.625	0.54	0.684	0.662
Avg PR	0.784	0.803	0.782	0.825	0.766
GPM	0.5967	0.6342	0.5838	0.6777	0.6582

Rule-based Classifiers

Under this classifier category, we utilized: (1) JRIP; (2) Decision Table classifiers and depicted the results in Table 4. It is observed that JRip showed superior performance than a decision tree in terms of threshold metrics, but a worse performance in terms of probability metrics and FPR, ROC, PR ranking metrics.

Table 4. Comparative results of Rule-based classifiers.

Rule-based Classifiers/Metrics	JRip	Decision Tree
CR	0.5555	0.3621
CPE	1.2974	1.104
RMS error	0.4209	0.4046
Avg DR	0.556	0.362
Avg FPR	0.112	0.094
Avg ROC	0.702	0.769
Avg FM	0.547	0.347
Avg PR	0.742	0.754
GPM	0.5198	0.4790

Bayes Classifiers

Under this category, we utilized: (1) BayesNet and (2) Naïve Bayes classifiers. The results are shown in Table 5. It is observed that Naïve Bayes shows superior performance than BayesNet in terms of threshold metrics and DR and PR-based ranking metrics. On the whole in this category, Naïve Bayes outperformed BayesNet classifier as indicated by GPM.

Table 5. Comparative results of Bayes classifiers.

Bayes/Metrics	Naïve Bayes	BayesNet
CR	0.5656	0.5109
CPE	0.8671	0.9967
RMS error	0.4136	0.4111
Avg DR	0.566	0.511
Avg FPR	0.233	0.185
Avg ROC	0.746	0.791
Avg FM	0.561	0.49
Avg PR	0.645	0.597
GPM	0.5625	0.5243

Functions Classifiers

Under this category, we utilized: (1) MLP; (2) SMO; (3) RBF Network and (4) LibSVM classifier (Chang and Lin 2011). The results are depicted in Table 6.

Table 6. Comparative results of Function-based classifiers.

Functions/ Metrics	SMO	MLP	RBF Network	LibSVM
CR	0.566	0.5797	0.4589	0.5779
CPE	0.8867	0.8682	0.9697	0.887
RMS error	0.377	0.3795	0.4051	0.4109
Avg DR	0.566	0.58	0.459	0.578
Avg FPR	0.205	0.202	0.239	0.204
Avg ROC	0.613	0.794	0.764	0.687
Avg FM	0.552	0.566	0.455	0.565
Avg PR	0.681	0.673	0.583	0.693
GPM	0.5577	0.5826	0.4986	0.5679

Lazy Classifiers

Under this category, we utilized: (1) IB1; (2) IBK and (3) KStar classifiers and show the results in Table 7. It is observed from the results that IBk shows a better performance than the other classifiers in this category in terms of threshold metrics, probability metrics and ranking metrics except PR. IBk gave the best performance in this category in terms of GPM.

Table 7. Comparative results of lazy classifiers.

Lazy/ Metrics	IBl	IBk	KStar
CR	0.5841	0.5868	0.5211
CPE	0.8328	0.8287	0.9541
RMS error	0.4079	0.4064	0.4244
Avg DR	0.584	0.587	0.521
Avg FPR	0.1	0.1	0.256
Avg ROC	0.742	0.818	0.589
Avg FM	0.587	0.589	0.474
Avg PR	0.748	0.75	0.617
GPM	0.6058	0.6152	0.5034

Meta Classifiers

Under this category, we utilized: (1) Bagging; (2) Boosting and (3) Random sub-space using Decision Tree-based J48 classifier and depicted the results in Table 8. It is observed from the results that bagged tree-J48 provides a better performance than the other classifiers in this category in terms of threshold metrics, probability metrics and ranking metrics. Bagged tree-J48 showed the best performance in this category in terms of GPM.

Table 8. Comparative results of meta classifiers.

Meta/ Metrics	Boosted Tree-J48	Bagged Tree-J48	Random Subspace-J48
CR	59.21	66.09	52.474
CPE	0.815	0.6941	0.8325
RMS error	0.3701	0.3498	0.3897
Avg DR	0.592	0.661	0.515
Avg FPR	0.056	0.047	0.064
Avg ROC	0.822	0.861	0.785
Avg FM	0.608	0.697	0.548
Avg PR	0.824	0.827	0.825
GPM	0.6410	0.6957	0.6035

Result Analysis

Best classifiers in various categories further can be compared based upon identified metrics as shown in Table 9.

Table 9. Categorywise comparative results of best classifiers.

		Threshold		Prob.		Ranking				
Category	Classifier	CR	Avg FM	CPE	RMSE	Avg DR	Avg FPR	Avg ROC	Avg PR	GPM
Tree based	J48	0.6485	0.684	0.716	0.373	0.648	0.049	0.792	0.825	0.6777
Rule based	JRip	0.5555	0.547	1.2974	0.4209	0.556	0.112	0.702	0.742	0.5198
Bayes	Naive Bayes	0.5656	0.561	0.8671	0.4136	0.566	0.233	0.746	0.645	0.5625
Functions	MLP	0.5797	0.566	0.8682	0.3795	0.58	0.202	0.794	0.673	0.5826
Lazy	IBk	0.5868	0.589	0.8287	0.4064	0.587	0.1	0.818	0.75	0.6152
Meta	Bagged Tree-J48	0.6609	0.697	0.6941	0.3498	0.661	0.047	0.861	0.827	0.6957

Bagged Tree-J48 classifier outperformed all the other classifiers by reporting optimal values of various metrics. J48 classifiers performed better than the other classifiers after bagged J48 in terms of threshold metrics, probability metrics, DR. Table 10 depicts the comparison of various best classifiers for different attack class categorywise.

Table 10. Attack classwise comparative results of best classifiers categorywise.

Attack Class	Category	Classifier	DR	FPR	ROC	FM	PR
Probe	Tree based	J48	0.962	0.111	0.89	0.651	0.495
	Rule based	JRip	0.899	0.019	0.986	0.87	0.843
	Bayes	Naïve bayes	0.718	0.008	0.907	0.802	0.908
	Functions	MLP	0.818	0.038	0.912	0.76	0.71
	Lazy	IBk	0.958	0.203	0.941	0.513	0.35
	Meta	Bagged Tree-J48	0.955	0.095	0.976	0.685	0.534
DoS	Tree based	J48	0.672	0.021	0.812	0.792	0.962
	Rule based	JRip	0.664	0.128	0.754	0.726	0.801
	Bayes	Naïve bayes	0.657	0.462	0.547	0.584	0.525
	Functions	MLP	0.652	0.391	0.766	0.605	0.565
	Lazy	IBk	0.631	0.124	0.798	0.705	0.799
	Meta	Bagged Tree-J48	0.672	0.017	0.851	0.793	0.968
U2R	Tree based	J48	0.05	0.08	0.43	0.01	0.003
	Rule based	JRip	0.184	0.001	0.674	0.262	0.452
	Bayes	Naive bayes	0.171	0.022	0.445	0.068	0.042
	Functions	MLP	0.101	0	0.663	0.171	0.561
	Lazy	IBk	0.061	0.002	0.639	0.092	0.182
	Meta	Bagged Tree-J48	0.053	0.08	0.523	0.007	0.004
R2L	Tree based	J48	0.455	0.022	0.722	0.607	0.911
	Rule based	JRip	0.161	0.018	0.514	0.269	0.817
	Bayes	Naïve bayes	0.326	0.039	0.933	0.465	0.807
	Functions	MLP	0.314	0.016	0.786	0.467	0.904
	Lazy	IBk	0.325	0.009	0.778	0.484	0.946
	Meta	Bagged Tree-J48	0.491	0.028	0.844	0.635	0.896

For probe attack class, Jrip classifier outperformed in terms of ROC and FM with DR of 89.9 per cent at 1.9 per cent FPR. Bagged tree-J48 classifiers detection probe attacks at a rate of 95.5 per cent at 9.5 per cent FPR. For DoS attack class, bagged tree-J48 outperformed other classifiers with the highest values of DR, ROC, FM, PR and lowest FPR. J48 tree-based classifier is at second position after bagged tree-J48 for this attack class. For U2R attack class, Jrip rule-based classifier outperformed other classifiers having maximum value for ROC, FM with DR of 18.4 per cent at 0.1 per cent FPR. For R2L attack class, Naïve Bayes classifier outperformed others in terms of ROC but shows less DR of 32.6 per cent at high value of 3.9 per cent FPR. Bagged Tree-J48 proved highest DR of 49.1 per cent at 2.8 per cent FPR.

Concluding Remarks and Future Work

In this empirical work, we performed a set of blind experiments of supervised classifiers on benchmarked KDD cup 1999 dataset. The dataset consists of connection record data comprising the data from network headers, network flows, logs and system events. Network flows, logs and system events, etc., generate big data. Big data analytics using AI-based techniques can correlate multiple information sources into a coherent view, identify intrusive activities and finally achieve effective and efficient intrusion detection.

The training and test dataset used in our experiments are a subset of KDD cup 1999 after preprocessing and removing all irrelevant and redundant features. The main objective of this work is to analyze the most commonly-used AI-based supervised classifiers for intrusion detection. The classifiers belong to different categories, viz: Tree based, Rule based, Functions, Bayes, Lazy and Meta classifiers.

From empirical results, it can be concluded that bagged tree-J48 classifier is best and stable classifier for organizations concerned with overall correct classification of malicious traffic with minimum cost for example, FPR and maximum ROC. Further, it is also suggested that rule-based JRip and Bagged Tree-J48 for probe, Bagged tree-J48 for DoS, JRip for U2R and Naïve Bayes, bagged tree-J48 and neural-network based MLP for R2L attack class can be utilized for detection of various attack classes.

Although the supervised classifiers are used directly for intrusion detection, our results prove that a single classifier cannot detect all the attack classes efficiently. Accordingly, a set of classifiers might be used to detect the attack belonging to different classes.

One more point can be highlighted in terms of detecting the attack classes. First, all supervised classifiers reported poor results for detection U2R and R2L attack classes. Similar results are also reported by (Subhnani and Serpen 2003). This might possibly be due to insufficient number of training instances in comparison to other attack classes. We need more analysis to explain these poor results.

These experiments and their results provide reliable guidelines for future research in applying supervised classifiers in field of intrusion detection and expose some new avenues of research. Further research is required to explore more supervised classifiers for intrusion detection. This fact is supported in our experiments that overall classification rate of 66.09 per cent is for best classifier in different categories. Furthermore, the overall performance of supervised classifiers might be enhanced by combining the strengths of individual classifiers. These hybrid models may be supported by voting type strategies that keep the strengths and weaknesses of individual classifiers into consideration.

For improving the performance of classifiers for detection of U2R and R2L attack classes, more informative features can be explored in future.

References

Amor, N. B., S. Benferhat and Z. Elouedi. (2004). Naive bayes vs. decision trees in intrusion detection systems. pp. 420–424. *In*: ACM Symposium on Applied Computing (SAC '04).

Caruana, R. and A. Niculescu-Mizil. (2004). Data mining in metric space: an empirical analysis of supervised learning performance criteria. KDD '04 Proceedings of the Tenth ACM SIGKDD International Conference on Knowledge Discovery and Data Mining, ISBN:1-58113-888-1.

Chang, C. C. and C. J. Lin. (2011). LIBSVM: A Library for Support Vector Machines, Available: http://www.csie.ntu.edu.tw/~cjlin/libsvm. Last accessed 20 January 2017.

Elkan, C. (2000). Results of the KDD'99 Classifier Learning, SIGKDD Explorations, ACM SIGKDD.

Gharibian, F. and A. A. Ghorbani. (2007). Comparative study of supervised machine learning techniques for intrusion detection. pp. 350–358. *In*: proceeding of Fifth Annual Conference on Communication Networks and Services Research (CNSR'07).

Halme, L. R. and R. K. Bauer. (1995). AINT Misbehaving: A taxonomy of anti-intrusion techniques. Proceedings of the 18th National Information Systems Security Conference. Baltimore, MD.

Hernandez-Pereira, E., J. A. Suarez-Romero, O. Fontenla-Romero and A. Alonso-Betanzos. (2009). Conversion methods for symbolic features: A comparison applied to an intrusion detection problem. Expert Systems with Applications 36: 10612–10617.

IBM. (2011). SPSS Statistics Software Package, available: www.spss.com/statistics/. Last accessed 20 January 2017.

Internet 2. (2010). Internet 2 Research Network. Available: http://www.internet2.edu/. Last accessed 20 January 2017.

KDD99, KDD cup 1999 data. (1999). Available: http://kdd.ics.uci.edu/databases/kddcup99/kddcup99.html.

Kumar, G., K. Kumar and M. Sachdeva. (2010a). An empirical comparative analysis of feature reduction methods for intrusion detection. International Journal of Information and Telecommunication 1: 44–51.

Kumar, G., K. Kumar and M. Sachdeva. (2010b). The use of artificial intelligence-based techniques for intrusion detection—A review. Artificial Intelligence Review 34(4): 369–387.

Kumar, G. and K. Kumar. (2012). An information theoretic approach for feature selection. Security and Communication Networks 5: 178–185.

McHugh, J. (2000). Testing intrusion detection systems: A critique of the 1998 and 1999 darpa intrusion detection system evaluations as performed by Lincoln Laboratory. ACM Transactions on Information and System Security 3(4): 262–294.

MIT. (2001). Lincoln Laboratory, 1999 DARPA intrusion detection evaluation design and procedure, DARPA Technical Report, Feb 2001.

Mukkamala, S. and H. S. Andrew. (2004). A comparative study of techniques for intrusion detection. *In*: Proceedings of the 15th IEEE International Conference on Tools with Artificial Intelligence (ICTAI'03).

Mukkamala, S., A. Sung and A. Abraham. (2005). Intrusion detection using an ensemble of intelligent paradigms. Journal of Network and Computer Applications 28(2): 167–182.

Pan, Z. S., S. C. Chen, G. Hu and D. Q. Zhang. (2003). Hybrid neural network and C4.5 for misuse detection. *In*: International Conference on Machine Learning and Cybernetics 4: 2463–2467.

Panda, M. and M. R. Patra. (2008). A comparative study of data mining algorithms for network intrusion detection. pp. 504–507. *In*: Proceedings of First International Conference on Emerging Trends in Engineering and Technology. IEEE Computer Society.

Ponce, M. C. (2004). Intrusion Detection System with Artificial Intelligence. FIST Conference—June 2004 Edition-1/28 Universidad Pontificia Comillas de Madrid.

Sabhnani, M. and G. Serpen. (2003). Application of machine learning algorithms to KDD intrusion detection dataset within misuse detection context. pp. 623–630. *In*: International Conference on Machine Learning: Models, Technologies and Applications, New York.

Shon, T. and J. Moon. (2007). A hybrid machine learning approach to network anomaly detection. Information Sciences 177: 3799–3821.

Singh, P. (2009). Comparing the effectiveness of Machine Learning Algorithms for defect prediction. International Journal of Information Technology and Knowledge Management 2(2): 481–483.

Sperotto, A. (2010). Flow-based Intrusion Detection. CTIT Ph.D.-thesis Series No. 10-180, Centre for Telematics and Information Technology, University of Twente.

Tavallaee, M., E. Bagheri, W. Lu and A. A. Ghorbani. (2009). A detailed analysis of the KDD CUP 99 data set. *In*: Proceedings of IEEE Symposium on Computational Intelligence in Security and Defense Applications (CISDA).

Tsai, C. F., Y. F. Hsu, C. Y. Lin and W. Y. Lin. (2009). Intrusion detection by machine learning: A review. Expert Systems with Applications 36(10): 11994–12000.

Weka. (2011). An open source data mining software tool developed at University of Waikato, New Zealand, Available: http://www.cs.waikato.ac.nz/ml/weka. Last accessed 20 January 2017.

Witten, I. H. and E. Frank. (2005). Data Mining-Practical Machine Learning Tools and Techniques, Second Edition, Morgan Kaufmann. An imprint of Elsevier. ISBN 0-12-088407-0.

Xu, X. (2006). Adaptive intrusion detection-based on machine learning: Feature extraction, classifier construction and sequential pattern prediction. International Journal of Web Services Practices 2(1-2): 49–58.

Zhang, J. and M. Zulkernine. (2005). Network intrusion detection using random forests. pp. 53–61. *In*: Proceedings of Third Annual Conference on Privacy, Security and Trust.

APPENDICES

Appendix A (Performance Metrics)

a. Classification rate (CR): It is defined as the ratio of correctly classified instances and total number of instances:

$$CR = \frac{Correctly\ classified\ instances}{Total\ Number\ of\ instances}$$

$$= \frac{TP+TN}{TP+TN+FP+FN}$$

b. Cost per example (CPE): It is computed as follows

$$CPE = \frac{1}{N}\sum_{i=1}^{5}\sum_{j=1}^{5} CM(i,j) * C(i,j)$$

where, CM corresponds to 5*5 confusion matrix, C corresponds to the 5*5 cost matrix and N represents the number of patterns tested. An entry in CM at row i and column j, CM(i,j) represents the number of mis-classified patterns, which originally belong to class i yet mistakenly identified as a member of class j. The cost matrix is defined in (Elkan 2000). A lower value for the CPE indicates a better classifier model.

c. Detection rate (DR): It is computed as the ratio between the number of correctly detected attacks and the total number of attacks.

$$CR = \frac{Correctly\ Detected\ Attacks}{Total\ Number\ of\ Attacks}$$

$$= \frac{TP}{TP+FN}$$

d. False positive rate (FPR): It is defined as ratio between number of normal instances detected as attack and total number of normal instances.

$$CR = \frac{Number\ of\ Normal\ Instances\ detected\ as\ Attack}{Total\ Number\ of\ Normal\ instances}$$

$$= \frac{FP}{FP+TN}$$

e. Precision (PR): Precision is the fraction of data instances predicted as positive that are actually positive.

f. F-measure (FM): For a given threshold, the FM is the harmonic mean of precision and recall at that threshold.

g. Area under ROC curve (ROC): ROC is a plot of sensitivity vs. 1-specificity for all possible thresholds. Sensitivity is the defined as P(Pred = positive | True = positive) and is approximated by the fraction of true positives that are predicted as positive (this is the same as recall). Specificity is P(Pred = negative | True = negative). It is approximated by the fraction of true negatives predicted as negatives. Area under the ROC curve is used as a summary statistic.

h. Root mean square error (RMSE): it measures how much predictions deviate from the true values. It is computed as

$$RMSE = \sqrt{\frac{1}{N} \sum (\text{Pr}\,edicitions - true\_values)^2}$$

CHAPTER **3**

Big Data Techniques in Social Network Analysis

B.K. Tripathy,* Sooraj T.R. and R.K. Mohanty

Introduction

A social network comprises of a finite set of actors, who are social entities. These entities can be discrete individuals, corporate or collective social units. They are related to each other through some relations, establishing some linkage among them. The social network has grown in popularity as it enables researchers to study not only social actors, but their social relationships. Moreover, many important aspects of societal networks and their study lead to the study of behavioral science. Scientific study of network data can reveal important behavior of the elements involved and the social trends. Important aspects of societal life are organized as networks. The importance of networks in society has put social network analysis at the forefront of social and behavioral science research. The presence of relational information is a critical and defining feature of a social network. Social network analysis is concerned with uncovering patterns in the connections between entities. It has been widely applied to organizational networks to classify the influence or popularity of individuals and to detect collusion and fraud. Social network analysis can also be applied to study disease transmission in communities, the functioning of computer networks and the emergent behavior of physical and biological systems. The network analyst would seek to model these

SCOPE, VIT University, Vellore-632014, Tamil Nadu.
Emails: soorajtr19@gmail.com; rknmohanty@gmail.com
* Corresponding author: tripathybk@vit.ac.in

relationships to depict the structure of a group. One can then study the impact of this structure on the functioning of the group and/or the influence of this structure on individuals within the group. The social network perspective thus has a distinctive orientation in which structures may be behavioral, social, political or economic.

Many evidences indicate the precious value of social network analysis in shedding light on social behavior, health and well-being of the general public. Social network analysis provides a formal, conceptual means for thinking about the social world. Freeman has argued that the methods of social network analysis provide formal statements about social properties and processes. Social network analysis thus allows a flexible set of concepts and methods with broad interdisciplinary appeal. It provides a formal, conceptual means for thinking about the social world. It is based on the assumption of the importance of relationships among interacting units. Of critical importance for the development of methods for social network analysis is the fact that the unit of analysis in network analysis is not the individual but an entity consisting of a collection of individuals, each of whom in turn is tied to a few, or many others, and so on. It attempts to solve analytical problems that are non-standard. The data are analyzed using social network methods and are quite different from the data typically encountered in social and behavioral sciences. However, social network analysis is explicitly interested in the interrelatedness of social units. The dependencies among the units are measured with structural variables. Theories that incorporate network ideas are distinguished by propositions about the relations among social units. Such theories argue that units are not acting independently from one another but, rather influence each other. Focusing on such structural variables opens up a different range of possibilities for and constraints on data analysis and model building. Instead of analyzing individual behaviors, attitudes and beliefs, social network analysis focuses its attention on social entities or actors in interaction with one another and on how these interactions constitute a framework or structure that can be studied and analysed in its own right.

The goal of social network analysis is to uncover hidden social patterns. The power of social network analysis has been shown much stronger than that of traditional methods which focus on analyzing the attributes of individual social actors. In social network analysis, the relationships and ties between social actors in a network are often regarded more important and informative than the attributes of individual social actors. Social network analysis approaches have been shown very useful in capturing and explaining many real-world phenomena, such as 'small world phenomenon'.

The rapid development of information technology over the last few decades has resulted in data growth on a massive scale. Users create content, such as blog posts, tweets, social-network interactions and photographs; servers continuously create activity logs; scientists create measurement data about the world we live in; and the internet, the ultimate repository of data, has become problematic with regard to scalability (Design for Scalability, Wikipedia 2018). This has necessitated the use of big data analysis techniques in the realm of social networks.

The usefulness of the study of big data can be gauged from the observations below and the use of big data analysis in different application areas are listed below.

- Edx (Online Education Platform) is a Harvard/MIT initiative for a free online-learning model; however, in return the platform collects a lot of data about students' experiences so that universities can offer a better experience on the campus.

- Recommender systems. The leaders in this field are Amazon and NetFlix. These companies can offer a better user experience by leveraging big data technologies.

- Urban planning development. Research shows (Bigdata-urban Planning, Montjoye de Yves-Alexandre et al. 2013) that aggregating data from a mobile network can identify the user's path during a time span so that business and traffic can be enhanced accordingly.

- Government. One of the best case studies is President Obama's campaign (Available at http://www.technologyreview.com) run by using big data technologies. By using cloud computing resources and the huge data that they have about voters, the campaign was able to reach more people to ask them to give their votes to Obama.

- Healthcare systems. Big data plays valuable a role in healthcare systems mostly in helping physicians to diagnose diseases quickly. An example is IBM Watson (Available at http://www.nytimes.com). Watson is currently trying to learn what is available in the literature about many diseases by using artificial intelligence techniques and then Watson will be able suggest the best available medicine in a matter of seconds.

Big Data Analysis

The importance of big data lies in how we utilize it as these types of data can be obtained from any source and their analysis can be done to attain cost reduction, time reduction, new product development and optimized offerings and smart decision making. The term 'big data' started to show up sparingly in the early 1990s and its prevalence and importance increased exponentially as years passed. Nowadays big data is often seen as integral to a company's data strategy. Big data has specific characteristics and properties that can help us understand both the challenges and advantages of big data initiatives.

Before going to describe the analysis techniques for big data, we shall see some of the important characteristics of these data sets. Since the origin, these data sets are characterized in terms of properties starting with the English alphabet V. It started with three Vs and this number is in the ever-increasing trend. Described below are seven Vs characterizing big data:

Volume

Volume is probably the best known characteristic of big data and this is no surprise, considering more than 90 per cent of all today's data was created in the past couple of years. The current amount of data can actually be quite staggering. Here are some examples:

- 300 hours of video are uploaded on YouTube every minute.
- An estimated 1.1 trillion photos were taken in 2016, and that number is projected to rise by 9 per cent in 2017. As the same photo usually has multiple instances stored across different devices, photo or document-sharing services as well as social media services, the total number of photos stored is also expected to grow from 3.9 trillion in 2016 to 4.7 trillion in 2017. In 2016 the estimated global mobile traffic amounted for 6.2 Exabyte per month. That's 6.2 billion gigabytes.

Velocity

Velocity refers to the speed with which data is being generated, produced, created, or refreshed. Sure, it sounds impressive that Facebook's data warehouse stores upwards of 300 petabytes of data, but the velocity at which new data is created should be taken into account. Facebook claims 600 terabytes of incoming data per day. Google alone processes on average more than '40,000 search queries every second', which roughly translates to more than 3.5 billion searches per day.

Variety

When it comes to big data, we don't only have to handle structured data but also semi-structured and mostly unstructured data as well. As can be deduced from the above examples, most big data seems to be unstructured, but besides audio, image, video files, social media updates, and other text formats there are also log files, click data, machine and sensor data, etc.

Variability

Variability in big data's context refers to a few different things. One is the number of inconsistencies in the data. These need to be found by anomaly and outlier detection methods in order to facilitate any meaningful analytics to occur.

Big data is also variable because of the multitude of data dimensions resulting from multiple disparate data types and sources. Variability can also refer to the inconsistent speed at which big data is loaded into your database.

Veracity

This is one of the unfortunate characteristics of big data. As any or all of the above properties increase, the veracity (confidence or trust in the data) drops. This is similar to, but not the same as, validity or volatility (see below). Veracity refers more to the provenance or reliability of the data source, its context, and how meaningful it is to the analysis based on it. For example, consider a data set of statistics on what people purchase at restaurants and these items' prices over the past five years. You might ask: Who created the source? What methodology did they follow in collecting the data? Were only certain cuisines or certain types of restaurants included? Did the data creators summarize the information? Has the information been edited or modified by anyone else?

Answers to these questions are necessary to determine the veracity of this information. Knowledge of the data's veracity in turn helps to better understand the risks associated with analysis and business decisions based on this particular data set.

Validity

Similar to veracity, validity refers to how accurate and correct the data is for its intended use. According to Forbes, an estimated 60 per cent of a data scientist's time is spent cleansing the data before being able to do any analysis. The benefit from big data analytics is only as good as its underlying data, so one needs to adopt good data governance practices to ensure consistent data quality, common definitions and metadata.

Value

Last, but arguably the most important of all, is value. The other characteristics of big data are meaningless if one is unable to derive business value from the data.

Substantial value can be found in big data, including understanding one's customers better, targeting them accordingly, optimizing processes and improving machine or business performance. One needs to understand the potential, along with the more challenging characteristics, before embarking on a big data strategy.

Big Data Tools

There are thousands of big data tools available. All of them are promising to save time, money and help to uncover never-before-seen business insight. In this section, we discuss some of the big data tools in detail.

Hadoop: Core of Big Data

Distributed File Systems (DFS) was not a new concept, but came from the need to share data from centralized data storage among different clients. DFS basically allows multiple clients to access files remotely on the file server for open(), close(), read(), and other operations. NFS (Network File System) is widely used as a distributed file system over a network. It was developed by Sun Microsystems as open protocol standards. NFS allows clients to access files remotely as they appear as they were stored directly to client storage (Fig. 1).

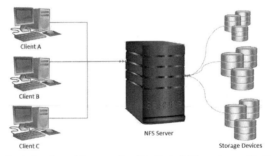

Fig. 1. A simple architecture for distributed file system with NFS.

Even though NFS is a powerful distributed file system, it has one main disadvantage—it stores all its volume data on one single machine. This means that the system has less reliability when the server goes down. Also, there is a storage limitation; the system can only store as much data as the machine can handle.

Today, NFS cannot fill the needs of big data. So, Hadoop was designed to store very large data on multiple nodes up to terabytes or even petabytes. So, what is Hadoop and how does it work? Hadoop (Tom 2011, Lin 2018) is an open source implementation of Google File System (GFS) developed by the Apache Project (Yahoo 2018). It was designed by Doug Cutting when he was working with Yahoo. Hadoop uses the Master/Slave model. The master is called Name-Node in the cluster and it manages the blocks of metadata and storage namespace and other requests that come from the clients to the file system. Name-Node also stores information about where those data blocks are stored in the cluster, whereas the slave works mainly as a data-node for storing data blocks as well as a task-tracker for MapReduce computation. There are two major components or layers in the Hadoop system—the HDFS layer and the MapReduce layer. Figure 2 below shows an abstract view of Hadoop cluster components.

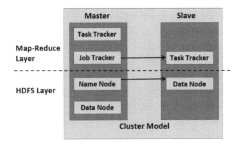

Fig. 2. Hadoop cluster components.

HDFS Layer

HDFS (Hadoop Distributed File System) is a robust distributed file system that was designed to store very large data—terabytes, or even petabytes of data. HDFS splits up large input files into small chunks or blocks. Each block has a fixed size—64 MB (default). These blocks are distributed and stored on different data nodes in the cluster.

MapReduce Layer

The MapReduce layer consists of JobTracker and TaskTracker. JobTracker is responsible for job assignments, task monitoring, job status and so on. TaskTracker, on the other hand, has responsibilities in running map and reduce tasks. To understand fully how this process works, the steps taken by the framework are as following:

- When the client submits a job, the JobTracker contacts Name-Node for data block locations.
- The number of map tasks is determined by how many blocks this specific input file has.
- The JobTracker accordingly assigns map tasks to the tasktrackers to run on each data block to produce what is called 'intermediate key-value pairs'. Map tasks store the output to the local file system of that specific node.

- When all Map tasks finish, the sorting, grouping and shuffle based on keys are done by the framework; similar keys merge together to become one key and a list of associative values. Then each reduce task consumes part of the shuffled data and output back to HDFS.

When all Reduce tasks finish, all intermediate data stored in temporary files are deleted upon the successful completion of the job. To understand this process more, a word-count example using MapReduce is explained in Fig. 3:

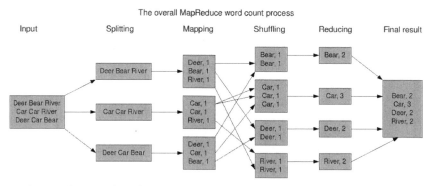

Fig. 3. A diagram of word-count example using Hadoop-MapReduce (http://www. confusedcoders.com).

MongoDB

MongoDB (http://www.mongodb.org) is schema-free, document-oriented, non-relational databases (NoSQL). It was written in C++ by 10gen Corporation as an open source project. MongoDB architecture has three core components:

- *Mongod* process handles data request and manages the underlying data format and data store.
- *Mongos* is the routing controller service for shard cluster. It handles incoming request from application layer in shards cluster and generates a response.
- *Mongo* is a JavaScript shell that helps the administrator to perform a wide range of query operations for testing, among others.

Figure 4 above shows simple standalone architecture and how MongoDB works. An administrator starts a *mongod* instance for handling data request and listens on '127.0.0.1' IP address if it is on a local machine (default port is 28017). Then *mongo* instance is started and interactively sends back and forth requests to *mongod* instance. MongoDB uses BSON (Binary Simple Object Notion), a binary encoded format of JSON. BSON supports the same data types that JSON supports. A BSON document

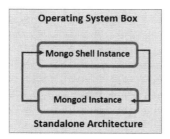

Fig. 4. Standalone architecture of MongoDB.

can store *strings, numbers, arrays, data objects* and even another BSON document.

MongoDB Data Modeling

In practical applications, schema design is a very important factor in any database management system. In MongoDB, as it has dynamic schema design, documents in real applications share a similar structure for good performance. There are patterns techniques required as follows:

Embedding

It is similar to de-normalization, which means that instead of having many one-to-one relationships among documents, by embedding those other documents into one gives a better read performance. In this case, a read operation can be done in a single query, for instance, student enrollment at a university. The university keeps records of the students' names, their education and their addresses. Each one is a separate document. However, by using the embedding technique, a read operation about a student's record can be done in one single query.

Referencing

Also called normalization, this is used in the case of one-to-many relationship among documents. For example, a hospital stores information about its doctors. Each doctor has many patients. The first scenario is to store an array object in the doctor's document that has all the patients' ids.

MongoDB supports atomicity on a single document level for write operations. For that reason, it's sometimes better to use an embedding data model to insure this property when it's required by the application. For instance, in the example about a student's record, atomic update can be done quickly. The document can be found based on the student id and through set expression—the subdocument 'education' also can be changed as well (Zhu et al. 2011). As in the example below, the status value was changed and another new field was added simultaneously.

There are two kinds of distributed architecture for the MongoDB. These are:

Replica-set cluster

MongoDB supports automatic failover in replica-set cluster. Replica-set architecture allows MongoDB to replicate its data across multiple nodes to insure high data availability (basically available). Each node runs its own *mongod* instance. The replica-set model is similar to the Master/Slave approach; however, Master/Slave doesn't support the automatic failover. When the Master goes down, an immediate intervention is required. Replica-set model has primary node (Master) and can handle up to 12 nodes in the cluster. All read-and-write operations are handled by the primary one and propagates any changes to all other secondary nodes (eventual consistency). In case of failover or if the primary is unreachable, the secondary members trigger an election. When a secondary gets majority votes, it becomes the primary. For example, suppose a replica-set (A) has three nodes, one primary and the other two secondary. Figure 5 shows a replica-set cluster with three members, one primary and the other two secondary.

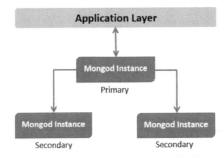

Fig. 5. A MongoDB replica-set architecture with three members.

Shard cluster

MongoDB is a horizontally scalable database. MongoDB allows sharding at a collection level (MongoDB Tutorial 2018).

Sharding is one of the more complex features provided by MongoDB and is a mechanism for scaling writes by distributing them across multiple shards. Each document contains an associated shard key field that decides on which shard the document lies.

A collection is partitioned to smaller chunks based on shard keys. A shard is a collection of documents within the shard key range. Furthermore, each shard can be partitioned to smaller chunks if the size of the shard exceeds 64 MB, also within the key range. The shard key is a similar field across all the documents within a collection. A shard key could be indexes or document ids.

To deploy a shard cluster, the following steps are adopted:

1. To enable a shard cluster on a database, three *Mongod* instances are required to work as a layer above the shard nodes. Each *configsvr* runs on its own machine. Those three *Mongod* store the chunks of metadata and that's where they are located.

2. At least three *mongos* instances run on a different machine each for reliability. *Mongos* instance cache and store the metadata of the *configsvr* and route all read and write operations to its shard container.

3. Shard containers can have either one *Mongod* instance or a replica-set of *Mongod* instances for high availability.

Figure 6 shows the architecture of a shard cluster. It has three *mongos* instances, three *configsvr* and multiple shard containers.

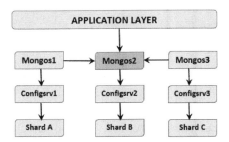

Fig. 6. An architecture of MongoDB shard cluster.

Exploring Big Data in Social Networks

What happens in 60 seconds? 168 millions e-mails, 694445 Google searches, 695000 Facebook status updates, 370000 Skype calls are made, 98000 tweets on Twitter, etc. The fundamental challenge in big data is not collecting the data; the main task is to making sense of it. Some of the challenges in online social networking research are explosive growth in size, complexity and unstructured data (Fay et al. 2006).

Enabled by various experimental methods: Observational studies, simulations, huge amount of data. It's big data, the vast sets of information gathered by researchers at companies like Facebook, Google and Microsoft from patterns of cell-phone calls, text messages and Internet clicks by millions of users around the world. Companies often refuse to make such information public, sometimes to protect the customer's privacy.

Computational sociology: A natural sciences approach to gather and analyze OSN data to study problems in sociology.

Social computing: An engineering approach that builds systems to support or leverage human social interactions and understand human behavior.

Data Mining Algorithm in Social Network Analysis

Data mining is an interactive process within which progress is defined by discovery through either automatic or manual methods. Businesses can learn from their transaction data more about the behavior of their customers and therefore can improve their business by exploiting this knowledge. Web usage information can be analyzed and exploited to optimize information access. Thus data mining generates novel, unsuspected interpretations of data (agrawal and Ramakrishnan 1994, Rajaraman and Ullman 2012). The main idea of data mining falls under two categories:

- Predictive data mining creates the model of the system from the given data.

- Descriptive data mining generates significant data sets from the existing data.

The aim of these above ideas is achieved by the following data mining techniques:

- **Characterization:** Characterization is used to generalize, summarize and possibly different data characteristics.

- **Classification:** Data classification is a process in which the given data is classified into different classes according to a classification model.

- **Regression:** This process is similar to classification; the major difference is that the object to be predicted is continuous rather than discrete.

- **Association:** In this process the association between the objects is found. It discovers the association between various data bases and the association between the attributes of single database.

- **Clustering:** Clustering involves grouping of data into several new classes such that it describes the data. It breaks large data set into smaller groups to make the designing and implementation process simple. The task of clustering is to maximize the similarity between the objects of classes and reduce the similarity between the classes.

- **Change Detection:** This method identifies the significant changes in data from the previously measured values.

- **Deviation Detection:** Deviation detection focusses on the major deviations between the actual values of the objects and their expected values. This method finds out the deviation according to the time as well the deviation among different subsets of data.

- **Link Analysis:** It traces the connections between the objects to develop models based on patterns in the relationships by applying graph theory techniques.
- **Sequential Pattern Mining:** This method involves discovery of the frequently occurring patterns in the data.

Social network is a heterogeneous and multi-relational data set. It can be represented by a graph, where vertices represent the objects or entities and edges represent the links or relationships between the objects. Both objects and links may have attributes. With the help of social networks, we can represent real time phenomena like phone calls, spreading of computer viruses, etc. These involve a lot of challenges from social network construction, such as information extraction and preprocessing to the data structures used for knowledge representation and storage.

A social network can be generically understood to be some kind of computer application which facilitates the creation or definition of social relations among people based on acquaintance, general interests, activities, professional interests, family and associative relations, and so on. Social networks can arise from information in sources, such as text, databases, sensor networks, communication systems (Whittaker et al. 2002) and social media. Finding and representing a social network from a data source can be a difficult problem. This challenge is due to many factors, including the ambiguity of human language, multiple aliases for the same user, incompatible representations of information and the ambiguity of relationships between individuals.

Some mining methods and algorithms will be introduced to describe the social networks. A method based on similarity measurement and inductive logic programming is useful here to analyze social networks. Moreover, some specific datasets are used to analyze the characters of social networks by Ucinet, which is a very powerful data mining software. Traditional data mining uses a 'property-value' table to represent data, which is shown as vector. Each dimension of vector corresponds to a value of conditional property. Social network data is a structured relational data and except for the properties of each node the more important thing is the link between the nodes. The links contain a lot of information. However, the vector form shows the independence of nodes. Therefore, to analyze the social network data, the relational data mining methods should be used. Social network analysis is a major application of relational data, whose development provides a more effective tool for social network analysis. The social network analysis pays attention to the link, which is a very important characteristic. From the aspect of data mining, the social network analysis is also called link mining (Ulrike Gretzel 2001). Through the mining of links, richer and more accurate information of instances can be obtained. At the same time, the link itself is often of concern to us as a source of information. For example, in certain circumstances, all the links are not observed. When a social network model is established, some very useful information can be analyzed from it.

Method based on inductive logic programming

Many algorithms of relational data mining are obtained from the ILP (inductive logic programming) (Freeman 2004). In order to study useful models from relational data, ILP method primarily uses the logic programming language. ILP is a crossover field of machine learning and logic programming, which is mainly concerned with finding new knowledge from data. Social network analysis is a major application of relational data, which provides a more effective tool for social network analysis.

ILP is an important method in learning about relationships, and construct the first order logic statements inductively from samples and background knowledge. Because it adopts logic as a representation, ILP overcomes two difficulties in traditional machine learning: limited expression of propositional logic and inability to provide background knowledge in the learning process. Furthermore, the result of ILP learning is easy to understand. ILP research focuses on the induction of relationship rule. In recent years, ILP research has been expanded to cover almost all the learning tasks, such as classification, regression, clustering and correlation analysis. Kings (Goh et al. 2003) uses the ILP method to classify the graph, whose representation is converted to relation representation. The predicates mentioned below are used to describe the graph:

Vertex (graphID, VertexID, VertexLabel, VertexAttributes)   (1)

Edge (graphID, VertexID1, VertexID2, BondLabel)   (2)

The ILP system can be used to find a good assumption in this space and has been used efficiently in mining subgroup and graph classification.

Method based on similarity measurement

Many data mining methods are based on similarity measurement. The definition of similarity is associated with the tasks. The best definition of similarity is likely to be different when the same data is used under different tasks. Sometimes it is difficult to choose an appropriate similarity measurement, especially when there are lots of attributes whose relationships are not clear regarding the aim and task. However, if given an appropriate similarity measure, such algorithm has a good intuitive explanation.

Similarity measure (Freeman 1979) is very useful in link prediction which is used to determine a link between two actors. In the social network G, the similarity measurement function for each pair of nodes, <x, y>, is given a possibility of a link: score (x, y). In some applications, the function can be seen as the topology structure of the network G, for each node x and y to calculate the degree of similarity between them. However, in some social network analysis, the weight is not calculated but the degree of similarity between nodes, is in order to do appropriate changes for a specific target. Some of these weights are based on neighborhood of node; others are based on the ensemble of all paths.

Now moving on to the weight based on the neighborhood, if two authors' colleagues have a large intersection, the possibility of their future cooperation will be greater than the same two who do not have same colleagues. Two people who have overlapping social circles have more probability to become friends. Starting from this intuitive observation, the probability that node x and node y contact each other in future is related to their neighborhood nodes, where $\Gamma(x)$ can be used to denote neighborhood of x in graph G. Some methods measure the intersection level of $\Gamma(x)$ and $\Gamma(y)$ as the probability of two node intersection.

Common neighbours (Carley 2003) use the number of neighbour's intersection as a measurement of intersection degree, which is a very straight idea. It is defined as:

$$score(x, y) = |\Gamma(x) \cap \Gamma(y)| \tag{3}$$

Jaccard coefficient refers to a similarity measure which is often used in information retrieval:

$$score = \frac{|\Gamma(x) \cap \Gamma(y)|}{|\Gamma(x) \cup \Gamma(y)|} \tag{4}$$

These two methods are simple counts, which treat all the neighbours equally, but Adamic/Adar method takes neighbours property into consideration, which weight the neighbours:

$$score(x, y) = \sum_{z \in \Gamma(x) \cap \Gamma(y)} \frac{1}{\log |\Gamma(z)|} \tag{5}$$

According to the node characters, as well as local or global structure of the networks, the importance of nodes can be judged. For example, node degree can simply be used as the important standard of local standards. Overall standards can use approach of eigenvectors to describe node importance which is related to the important nodes that are linked.

Specific interpretation of social network analysis

Social networks analysis is a set of norms and methods to analyze social network structure and its properties. It is also called structural analysis, because it mainly analyses structures and attributes of social relations constituted by different social units, such as individuals, groups, organizations and so on. Hence, social network is not only a set of technologies to analyse the structure and relations, but also a theoretical approach, which is called structural analysis idea.

Network analysis is to explore the deep structure which is a certain network mode hidden under the complex social system surface.

Basic principles of social network analysis

As a basic approach to research in social structure, there are several principles of social network analysis. First, relationship ties are often interactive and asymmetric; it is different in content and intensity. Second, relations link directly or indirectly to the network members; so we must analyse it under a larger the network structure context. Third, social ties' structure produces non-random networks, which generate network clusters, network boundaries and cross-correlation. Fourth, cross-correlation links network groups and individuals together. Fifth, asymmetrical ties and complex networks make the distribution of scarce resources unequal. Sixth, network produces cooperation and competition for the purpose of scarce resources. This structural methodology (Goh et al. 2003) means social science research should be targeting the social structure, rather than the individual. Through research network relationships, we can associate relationships between the individuals, 'micro' social networks and large-scale social systems of 'macro' structure together (Michael et al. 2007). Therefore, the British scholar, J. Scott, said: "Social network analysis has laid the foundation of new theories emerging of social structure."

While the researchers emphasize on these methodology of social structure, they use the concept of structure which is much different. In fact, in sociology, social structure is used at different levels. It can be used not only to illustrate the micro social interaction relation model, but also to explain the macro-social relations model. In other words, from the social role to the society, there are structural relationships.

Levels of Social Structure

In general, sociologists use the concept of social structure at the following levels:

i) *The level structure of the social role (micro structure)*: The most basic social relation is role relation. Role is often not singular, isolated, but is a form of role bundle. It reflects people's social position or status relationship, such as teacher-student.

ii) *The level structure of organization or group (middle structure)*: Refers to the relationship between the constituent elements of society. This structure is not reflected in the individual relationship between activities. Such an occupational structure reflects the social-professional status, the relationship with the resources and so on.

iii) *The level structure of the social system (macro structure)*: Refers to the social macro structure as a whole. For example, the class structure of society embodies the relationship between the major interest groups, or the characters of social system.

Therefore, the social structure has multiple meanings. However, from the new concept of structure, the social structure is the general form of social existence rather than a specific content.

Therefore, many sociologists have advocated that the study object of sociology should be the social relations rather than specific social individual, because individuals are different and changing, while only the relationship is relatively stable.

Data Analysis

Here software NetDraw (Girvan and Newman 2002) can be used to draw graphs, especially of the social network data. The NetDraw supports a VNA data format which is in ordinary text file format. They have three start sections—node data, node properties, and tie data. The three sections consist of the structure of VNA file. The first section, 'node data' contains variables that describe the actors in a network, i.e., it contains some actors' features which can be used to divide actors into different subgroups. The second section 'node properties' contains some of the actors' attributes which can be shown on the graph. The third section 'tie data' contains the relations between actors. It may include the link relation and strength relation. 'Zachary' data file (Brandes 2001) can be used in data analysis to show relationships among people in Zachary club. Different data samplings are extracted in this example. The ucinet can show a circle graph by using samplings. The nodes here are located at equal distances around a circle. Nodes that are highly connected are very easy to locate because of the density of lines. If transforming from attribute 'ID', the result is shown as a left part of Graph 1. K-cores, which is a definition of a group or subgroup to show the graph's subgroup can be used too. Nodes in the same group have same color. A K-core is a maximal group of actors, all of whom are connected to some number (*k*) of other members of the group. In this example, there are four different subgroups as shown in right part of Fig. 7.

Using in-degree statistics, ZACHE data is analysed by ucinet, as shown as Table 1. An actor's in-degree is the sum of connections from one actor to other actors, showing how many actors send information to the specific one. The in-degree is very meaningful because actors that receive much information from sources seem to be powerful and prestigious. But sometimes it may mean information overload or noise interference.

In a similar way, many tables and graphs can be made by using ucinet analyzing data. Many quotas which help people to analyze the social network can be found too. Link proportion shows proportion of each actor's link to another. Adjacency shows which two actors have one actor to share; out-degree shows which actor sends information to others; reachability shows connections which can be traced from the source to the specific actor. The distance is important to understand actors' differences and the probability that they have their position; the maximum flow shows the number of paths between two actors; a group shows how the network as a whole is likely to behave. The tree diagram shows how many actors are overlapped in different cliques.

Fig. 7. Circle and K-cores.

Table 1. Descriptive statistics in degree.

		1	2	3	4	5	6	7	8	9	10	11	12	13	14	15	16
1	Mean	0.49	0.27	0.30	0.18	0.09	0.12	0.12	0.12	0.15	0.06	0.09	0.03	0.06	0.15	0.06	0.06
2	Std. Dev.	0.50	0.45	0.46	0.39	0.29	0.33	0.33	0.33	0.36	0.24	0.29	0.17	0.24	0.36	0.24	0.24
3	Variance	0.25	0.20	0.21	0.15	0.08	0.11	0.11	0.11	0.13	0.06	0.08	0.03	0.06	0.13	0.06	0.06
4	Euc. Norm	4.00	3.00	3.16	2.45	1.73	2.00	2.00	2.00	2.24	1.41	1.73	1.00	1.41	2.24	1.41	1.41
5	MCSSQ	8.24	6.55	6.97	4.91	2.73	3.52	3.52	3.52	4.24	1.88	2.73	0.97	1.88	4.24	1.88	1.88
6	SSQ	16	9	10	6	3	4	4	4	5	2	3	1	2	5	2	2
7	Sum	16	9	10	6	3	4	4	4	5	2	3	1	2	5	2	2
8	Minimum	0	0	0	0	0	0	0	0	0	0	0	0	0	0	0	0
9	Maximum	1	1	1	1	1	1	1	1	1	1	1	1	1	1	1	1
10	No. of Obs.	33	33	33	33	33	33	33	33	33	33	33	33	33	33	33	33

Conclusion

Nowadays, big data has become very common as most of the datasets have some or other characteristics under this category of data set. Social networks are being used in day-to-day life sparingly as several application tools have become compulsions for humans. The data generated while dealing with these social network tools come under big data category. It is well known now that using social network data for deriving conclusions and predictions can only be possible through a scientific analysis. Since the data sets fall under big data category, any analysis technique for social networks can only be possible by using techniques developed so far for big data analysis. This chapter presents fundamentals of big data, social networks and social network analysis. Our main focus was on describing the social network analysis techniques, which can handle big data. It is however felt that this branch of investigation is still in its preliminary stage and a lot of research is needed to develop efficient social-network-analysis techniques with the focus on big data.

References

Agrawal Rakesh and S. Ramakrishnan. (1994). Fast Algorithms for Mining Association Rules. pp. 1–13.

Bigdata-urban Planning. Available: http://web.mit.edu/newsoffice/2013/de-anonymize-cellphone-data-0327.html.

Brandes, U. (2001). A faster algorithm for betweenness centrality [J]. Journal of Mathematical Sociology 25(2): 163–177.

Carley, K. (2003). Dynamic Network Analysis. In: Ron Breiger and Kathleen M. Carley (eds.). The Summary of the NRC Workshop on Social Network Modeling and Analysis [C]., National Research Council.

Chang Fay, Dean Jeffery, Ghemawat Sanjay, Hsieh C. Wilson and W. A. Deborah. (2006). Bigtable: A distributed storage system for structured data. OSDI 7: 1–14.

Design for scalability Available: http://en.wikipedia.org/wiki/Scalability.

Edx : Online Education Platform. Available: https://www.edx.org/.

Freeman, L. (1979). Centrality in social networks: Conceptual clarifications. Social Networks 10(1): 215–239.

Freeman, L. C. (2004). The development of social network analysis[M]. Vancouver: Empirical Press.

Girvan, M. and M. Newman. (2002). Community structure in social and biological networks [C]. USA: Proc. Natl. Acad. Sci., pp. 8271–8276.

Goh, K. I., E. Oh, B. Kahng et al. (2003). Betweenness centrality correlation in social networks [J]. Physical Review E 67(1-2): 017101-1-017101-4.

IBM Watson. Available: http://www.nytimes.com/2013/02/28/technology/ibm-exploring-new-feats-for-watson.html?pagewanted=all&_r=0.

Lin, J. Cloud9 using hadoop. Available: https://github.com/lintool.

Michael, M., Moreira E. José, D. Shiloach and W. W. Robert. (2007). Scale-up x scale-out: A case study using Nutch/Lucene. pp. 1–8.

MongoBD, http://www.mongodb.org/.

Montjoye de Yves-Alexandre, Hidalgo A. Ce´sar, Verleysen Michel and B. D. Vincent. (2013). Unique in the Crowd: The privacy bounds of human mobility. Scientific Reports, pp. 1–5.

Obama's campaign, available @ http://www.technologyreview.com/featuredstory/508856/obamas-data-techniques-will-rule-future-elections/.

Rajaraman, A. and J. D. Ullman. (2012). Mining of Massive Datasets. Cambridge University Press, pp. 1–310.

Shard Cluster Architecture. Available: http://docs.mongodb.org/ecosystem/tutorial/install-mongodb-on-amazon-ec2/.

Tom, W. (2011). Hadoop: The Definitive Guide, Second Edition ed.: O'Reilly Media, Inc.

Ulrike Gretzel. (2001). Social network analysis: Introduction and resource [EB/OL]. http://lrs.ed.uiuc.edu/tse-portal/analysis/socialnetwork-analysis/.

Wei-ping Zhu, Ming-xin Li and H. Chen. (2011). Using MongoDB to Implement Textbook Management System instead of MySQL. IEEE, pp. 303–305.

Whittaker, S., Q. Jones and L. Terveen. (2002). Contact management: Identifying contacts to support long term communication[C]. Proceedings of Conference on Computer Supported Cooperative Work, New York: ACM Press, pp. 216–225.

WordCount-example. Available: http://www.confusedcoders.com/.

Yahoo tutorial-custom data type. Available: http://developer.yahoo.com/hadoop/tutorial/module5.html#types.

CHAPTER 4

Analysis of Deep Learning Tools and Applications in e-Healthcare

Rojalina Priyadarshini, Rabindra K. Barik and
*Brojo Kishore Mishra**

Introduction

Machine learning (ML) is an area dealing with the behaviour of functional relationship among data without having thorough knowledge on the characteristics exhibiting in data. These techniques are applied to derive the functional relationship between data which do not contain their precise definitions. In real cases, the datasets, which are available, do not have a precise definition and functional co-relation. So, ML algorithms are used appropriately for those types of data which can infer the useful features of the data and their functional co-relationship for use in problems of classification, clustering and prediction, etc. Deep learning (DL) is a subfield of ML which is built on the algorithms inspired by the structure and utility of the human brain. This study is coherent in artificial neural network (ANN) and can be called advanced ANN having a higher set of problem-solving capability and utility. DL refers to a special type of ANN which is constituted by multiple numbers of hidden layers. The

C.V. Raman College of Engineering, Bhubaneswar, India.
Emails: priyadarshini.rojalina@gmail.com; rabindra.mnnit@gmail.com;
* Corresponding author: brojokishoremishra@gmail.com

hidden layers are responsible for handling nonlinearity and complexity of underlying data. A normal traditional neural network (NN) contains many simple, connected processing elements known as neurons, where every neuron is capable of generating some activated real-value signals upon processing the input. Input neurons get activated through the perception of the surrounding while the neighboring neurons get activated by the connected weights of previously active neurons. Out of these neurons, some may sway the environment by triggering actions, whereas some may not. A neuron learns the means if it finds the correct weight that makes the NN exhibit the desired behavior. Depending on the complication of the problem and how the neurons are connected, such behavior may need a series of computations, where each computation is done on a layer and the output of that corresponding layer is used as input for the subsequent layer. Different activation functions may be used to trigger the output of a particular layer. Deep Learning involves many subsequent layers, unlike the shallow NN.

These are mainly useful in giving training to network exhibiting complex co-relational functions and behavior and are mostly suitable for large data sets. Thus they are very successfully applied to big data domain. In this contemporary era, data is highly scalable and growing exponentially. There is a parallel growth in data reproduction, availability and access and abundance of high-speed computers. So the need of the hour is to develop high-speed large neural networks which are scalable to deal with a plethora amount of big data. The input to the conventional shallow-learning algorithms are known features from the data, but DL algorithms posses the potential to extract the representative features from the raw data, whose sources are social media, transactional data, business houses, etc.

In most applications of data, like classification, clustering and prediction, the common work flows for machine algorithms are built on some basic steps like: (i) data cleaning (ii) pre-processing (iii) training which build the learning model and (iv) testing which will evaluate the built model. For these problems the input space contains a set of input data sample x_i ($i = 1,2,...,n$; where n is the number of input instances which is a vector of 'm' dimension where m is the number of features) and output y_i. In supervised algorithms the artificial model learns to derive a mapping function $f(x) = y$ from a prior known pair of x_i, y_i. After the model is duly trained on the basis of known input-output pairs (x_i, y_i), it is evaluated against the unknown samples to infer the target output which is called testing.

In case of unsupervised model, the model has to explore the pattern exhibited in input x_i to find out y_i. In the training phase, output y_i is not provided in the model; it has to find out itself. Clustering, a principal component of analysis, is an example of unsupervised learning.

Deriving feature sets x_i to build an input dataset is very time-consuming and problem-specific. These are fewer and difficult to create for every problem. This is the reason for unsuitability of supervised and unsupervised algorithm in all the problems. But deducting the principal feature from a dataset is directly associated with the performance of a learning algorithm. If the features are correctly obtained from the data, then learning will take less time and the model shows a higher level of accuracy. On the other hand, for high dimensional data containing big data components, it is almost impossible to find out all the informative and relevant features from the raw data. This is a major limitation of using conventional machine-learning algorithms in big data possessing high-dimensional feature data. This limitation is overcome by the emergence of deep neural network DNN (Liu et al. 2017, Hubel et al. 1962). DNN has many layers contains more than one hidden layer. The input layer takes the high-dimensional raw data and extracts the relevant representative features from them. These abstract features are regenerated in subsequent hidden layers to incorporate the complex and nonlinear functions exhibited in high-dimensional big data. The potential of deep learning algorithms to handle high-dimensional raw data makes them popular in the field of data analysis as they deal with data having components like volume, variety, veracity and versatility of data.

Types of Deep Learning Network

The deep neural networks are categorized into different types according to their architecture and learning types. The broad classes include Convolutional Neural Network, Restricted Boltzmann's Machine, Deep Belief Network, Recurrent Network and AutoEncoders as depicted in Fig. 1.

Due to variations in methodology, there are several architectures proposed and used in DL. Two of the most common architectures are: Convolutional Neural Network (CNN) and Restricted Boltzman's

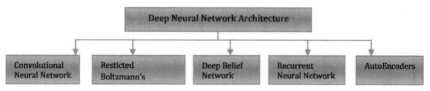

Fig. 1. Architecture variants of deep neural network.

Machine (RBM). In CNN, a set of infused feed-forward layers are present containing convolutional filters (Liu et al. 2017, Hubel et al. 1962). Every layer in the network generates a highly abstract set of features, which are further regenerated in the subsequent features. The process is similar to the architecture of the visual cortex which absorbs visual information in the form of receptive fields. Their variations can be Deep Belief Network (DBN) and AutoEncoders (AE). The architectural details are discussed below.

Deep Belief Network

DBN are composed of more than one layer of hidden units where connections exist between layers but there is no connection in between the nodes within each of the layers. The input data sets can be trained in unsupervised manner to reconstruct the inputs or to detect the features. Then a supervized algorithm can be used to do classification and prediction. So these can be seen as a simple unsupervised network like RBM or AE, where the hidden layer of the sub-network is worked as a visible layer to the subsequent layer. DBN can be visualized as a stack of RBM. The architectural model of DBN is given in Fig. 2, where it can be noticed that every two adjacent layers are nothing but one RBN. The visible nodes of one layer are connected to the hidden nodes of the subsequent layer. The lower nodes are trained first and then the higher nodes are trained in a sequential manner (Deng 2012, Bengio 2009). Training happens sequentially from down to top of the network. The training process of a DBN is carried out through a couple of stages: the first one is pre-training stage which is followed by a fine-tuning stage. During the former stage, unsupervized learning-based training is performed from bottom-up layer to do feature extraction while in the later stage, supervized learning based on top-down algorithm is applied to adjust several parameters of the network.

Hidden layer h$_2$ weight w$_2$

Hidden layer h$_1$

Visible nodes: V

weight w$_1$

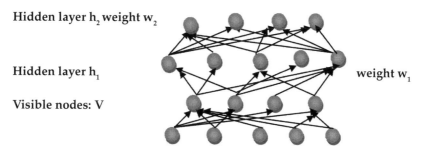

Fig. 2. Schematic diagram of Deep Belief Network.

AutoEncoder

AE is a variant of ANN and is known as autoassociator or Diabolo network designed to convert inputs to outputs with minimum quantity of noise and distortion. It uses an unsupervised learning algorithm to codify the input dataset. The main applications of these type are used in dimensionality reduction (Bengio 2009, Hinton and Salakhutdinov 2006) in the field of information processing. AE, in simple form, are analogous to the multilayer perceptron which contains an input layer and output layer and some hidden layers in between them. The number of nodes in the input and output layer are same, so that the inputs can be reconstructed rather than being predicted. This is a sort of unsupervized learning scheme used to reconstruct the outputs from the input. An AE has two processes—Coding and Decoding. In coding, it transforms the input vector 'i' to a concealed depiction 'h' applying a weight matrix ω, followed by decoding process, where AE gets back 'h' to the original form to get 'i' associated to another weight matrix ω'. Theoretically, ω' should be the transpose of ω. In this way, it learns to transforms the input into a separate representation by which the input is recreated out of the same representation. Measures are taken to minimize the average error rate in reconstruction and some means are followed to minimize the deviation between i and i'. The mean square error is a standard method to check the recreation accuracy (Hinton and Salakhutdinov 2006, Ling et al. 2015).

During the training process of an AE, in the first stage, the network learns features applying unsupervized learning and in the later phase, the network is finetuned where supervized learning algorithms are used to minimize the deviation rate between i and i'. So, through unsupervised learning, features are extracted from the raw input and the network is trained by the extracted features. Then, by supervized learning algorithms, some network parameters are optimized by back propagating the error between the i' and i. The conceptual level schematic representation of AE is given in Fig. 3.

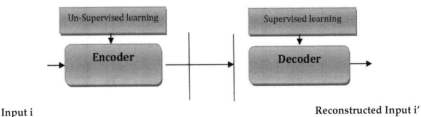

Fig. 3. Conceptual architecture of AutoEncoder.

Convolutional Neural Networks

CNN are a special type of neural network which are similar to feed forward artificial neural network where the connectivity of neurons are inspired by the animal's visual cortex. Bubel and Wise in 1960 (Hubel and Wiesel 1962, Van den Oord et al. 2013) revealed the concept of the receptive field. They are constituted by neurons containing adaptable weights and biases. They found that the cells present in an animal's visual cortex are arranged as overlapping sub-regions of visual fields. The images are transformed hierarchically. The response of each neuron to stimuli in respective visual fields can mathematically be represented by convolution operations. The response in case of CNN is inspired by this phenomenon, but convolution is substituted by normal matrix multiplications. These networks are well suited to processing of two-dimensional data having grid representations, e.g., for image and video processing (Arel et al. 2010, Collobert and Weston 2008). The images do not need to transform into any other domain; they can directly be imported to the network which does not require any feature extraction. So, a minimum amount of pre-processing is required. With the emergence of GPU, accelerated computing is used to train CNN in a faster and efficient way. These are successfully applied in the field of handwriting recognition, recommender systems (Van den Oord et al. 2013, LeCun et al. 2015) and natural language processing (Collobert and Weston 2008).

In a CNN the number of intermediate layers are more than one and it contains two different types of layers—a convolution layer (c-layer) and sub-sampling layer (s-layer) (Deng 2012, LeCun et al. 2015, Krizhevsky et al. 2012) as depicted in Fig. 4. There are alternative C-layers and S-layers present in between the input and output layers of CNN. The C-layers contain convolutional filters that are learnable (called also kernels containing tiny receptive fields) and which produce feature maps from the input images for the first c-layer. The connection weights, associated with the filters, are passed through a sigmoid function which generates added features for first s-layer. This process repeats for the subsequent layers. These feature maps are represented in the form of s-layer. At last the features are trained to obtain the output (Witten et al. 2016). There are three main components of learning in CNN—sparse interaction, parameter sharing and equi-variant representation (Goodfellow et al. 2016, Hinton et al. 2006). In traditional NN, the relation between the input and output neurons are mapped by matrix multiplication operations, but in CNN, the intensive matrix multiplication is reduced by making the interaction sparse.

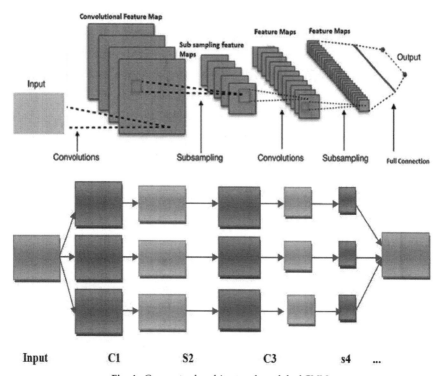

Fig. 4. Conceptual architectural model of CNN.

The main objective of a CNN is to identify the higher-order features in the data through convolutions. They are best fit to be applied in object recognition with images and sound. They are able to recognise faces, individuals, street symbols and many other aspects of visual data. CNNs are also capable to perform text analysis through optical character recognition. So they are equally useful at analyzing sound.

Restricted Boltzmann's Machines

RBMs, based on a sound theory of Boltzmann's machine, was first proposed by Smolen and was made famous by Hinton due to re-insight into the work (Yu and Deng 2011). RBMs are applied to include stochasticity into ANN models which can learn the probability distribution among inputs. RBMs are constituted of a variant of Boltzmann machines (BMs). BMs resemble NNs having stochastic processing units connected bi-directionally. But it is hard to learn the characteristics of an anonymous probability distribution. RBMs are used to simplify the topology of the network by increasing the efficiency of the model. In an RBM, the neurons are arranged in terms of a bipartite graph in which it is seen that there is no connection between the neurons of the same layer but there is full connection in between the nodes of different layers (Deng 2012). So the neurons are arranged into two groups—visible unit and hidden unit. The schematic representation is presented in Fig. 5.

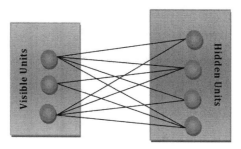

Fig. 5. Schematic diagram of RBM.

Recurrent Neural Network

RNN (Williams and Zipser 1989) is an NN that includes hidden nodes capable of analyzing streams of sequential data. This is vital in applications where the output depends on the previous calculations, such as the study of text, speech and synthesis of DNA and protein sequences. The RNN takes the inputs with training samples that possess a strong correlation and a meaningful representation to maintain order about the happenings of all the previous steps. The outcome obtained by the network at time $t-1$ affects the choice at time t. Like this, RNNs make use of two sources of input—the present input with the recent past, to provide the output of the new data. For this reason, it is often said that RNNs have memory. Although the RNN is a simple and powerful model, it suffers from vanishing gradient and exploding gradient problems as described in Bengio et al. (1994).

Recurrent networks are differed from feed-forward networks by one of their significant characteristics. They have feedback loops connected to their past decisions since their outputs are fed as their own input again repeatedly. It is considered that recurrent networks have memory. There exist a purpose for adding memory to neural networks. For example, protein data are sequential data correlated with each other and dependent. For this sort of data, where the information is contained within them, recurrent networks use this information to perform tasks that feed-forward networks are not able to do. Such sequential information is conserved in the recurrent network's hidden state, which manages to span many time steps. It is analogous to human memory that circulates invisibly within a body, affecting our behaviour without revealing its full shape. Information moves in the hidden states of recurrent networks.

Activation Functions Used in Deep Learning

Activation functions are used to bring in nonlinearity to deep neural network models, by which deep learning models are able to learn nonlinear prediction limits. The two major types of activation functions used are—logistic shape or rectifier shape.

Logistic Shape

Logistic functions resemble an 'S'-shaped curve. These activation functions are differentiable. The reason behind their usefulness is that they can force any input to be bounded between two outputs.

Sigmoid: The sigmoid function is given by Equation 1. Basically it is used for binary classification problems, where '*x*' is the input.

$$f(x) = \frac{1}{1 + e^{-x}} \tag{1}$$

The limitations of the activation function are given below:

- It suffers from vanishing gradient problem.
- Its output is not centred towards zero. It makes the gradient updates go too far in different directions. $0 <$ output < 1, and it makes optimization harder.
- Sigmoidal functions show slow convergence (Kaiming et al. 2015, http://www.deeplearning.net).

The shape of a sigmoidal curve is shown in Fig. 6.

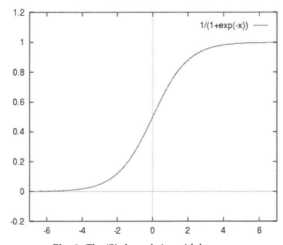

Fig. 6. The 'S'-shaped sigmoidal curve.

Softmax: The softmax function looks similar to the sigmoid function. When trying to predict two classes, the softmax function is equivalent to using the sigmoid activation function. When the outputs are more than two, then sigmoid does not help much. On the other hand, the softmax function squashes the outputs of every unit lying between 0 and 1, just like a sigmoid function. But in addition to that, it also divides each output in such a manner that the total sum of the outputs is equal to 1. The output of the softmax function is equivalent to a categorical probability distribution, conveying that the probability of any of the classes is true.

Mathematically the softmax function is presented in Equation 2, where x is a vector of the inputs to the output layer. And j is the index of the output units, starting from 1 to k; $j = 1, 2, ..., K$.

$$\sigma(x)j = \frac{e^{xj}}{\sum_{x=1}^{k} e^{xk}} \qquad (2)$$

Softmax function is applied to convey probabilities when there has been more than one output. Then it generates a probability distribution of outputs. It is useful in finding the most probable occurrence of output among all the other generated outputs.

Hard Sigmoid: Instead of calculating the sigmoid as defined above, an approximation of the sigmoid can be used. When trained, the hard sigmoid can generate similar outputs to the sigmoid and is much easier computationally to calculate. It is defined as a piecewise function that is either 0 slope when x is < -2.5 or $x > 2.5$ and has a 0.2 slope when x is $-2.5 < x < 2.5$. Because the hard sigmoid is easier for a computer to calculate, each iteration of learning is faster than with a traditional sigmoid. If time to train the model is a constraint, this activation function could lead to a smarter model (Bengio et al. 1994, http://www.deeplearning.net).

Rectifier Shape

Rectifier functions are characterized by having slopes of 0 or approaching 0 when the input is less than 0 and having slopes of 1 or close to 1 when the input is greater than or equal to 0. One of their benefits over logistic functions is that the output is not bounded between two asymptotes. Therefore, a rectifier can generate output values in the range of $[0, \infty)$, which is why rectifiers have become popular in solving the 'vanishing gradient problem' described earlier.

ReLU: ReLUs are very commonly used as activation functions in the hidden layers of a neural network. Hidden layers are the middle layers residing in between the output layer and they are not the first or last layers in the network. The simple mathematical function is shown in Equation 3 (http://www.deeplearning.net). A ReLu function outputs to 0 if the input is less than 0 and raw output is otherwise. That is, if the input is greater than 0, the output is equal to the input. ReLU's machinery is more like a real neuron in our body. Its curve is shown below in Fig. 7.

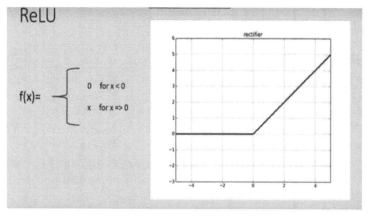

Fig. 7. ReLu function.

$$f(x) = \max(x, 0) \tag{3}$$

Leaky ReLU: In a standard ReLU, the slope of the activation function for input value of less than 0 is 0. However, in a Leaky ReLU, we can give x a small positive slope (let's call it the constant C), so that the network learns to move away from the negative x values.

$$f(x) = (Cx)(x < 0) + (x)(x >= 0)$$

The Leaky ReLU will have a slope of 1 when $x > 0$ and have a slope of C when $x < 0$. This allows the network to continue learning and avoid the trap of the network from becoming very sparse.

The only dissimilarity between ReLu and Leaky ReLu is that in Leaky ReLu, it does not completely vanish the negative component; it just lowers its magnitude as shown in Fig. 8 (http://nptel.ac.in/courses).

PReLU: PReLU is very similar to the Leaky ReLU function because the model uses a value to multiply by x when $x < 0$. But instead of just guessing a value to multiply x when $x < 0$, we train a parameter of our model to find the optimal value to multiply by x.

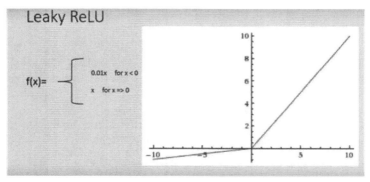

Fig. 8. Leaky ReLu function.

Comparison of Different Learning Architecture

Types	Description
Deep Neural Network	These are made up of more than two hidden layers. These are widely used in classification and regression problems. In most of the cases, labelled data are used for training. The training process may be slow for a large amount of training data.
AutoEncoders	These are specially designed for feature extraction and dimensionality reduction problem. They contain the same number of input and output nodes. Their focus is to recreate the input instances. They do not need labelled data for training, but they may require pre-training. One limitation of these types is that they may suffer from vanishing gradient problem.
Deep Belief Network	These are composed of subnetworks of many RBMs, but here the hidden layer of each subnetworks is visible to the next. Both supervised and unsupervised training is allowed in this type of network.
Restricted Boltzmann's Machine	These are based on Boltzmann's machine. They have undirected connections among all the layers. The neurons are connected like bipartite graph. They use stochastic maximum likelihood algorithms to impose stochasticity in NN.
Recurrent Neural Network	These are good at handling a stream of data. They are useful for problems where the output depends on previous calculations. They can memorise sequential events and are used to handle sequential data like genome data, protein data, etc. These networks may also suffer from vanishing gradient problem.
Convolutional Neural Network	These are good at handling image data. Herein every hidden layer, there are convolutional filters present. These are encouraged by the neurobiological model of visual cortex. They work efficiently for a huge dataset of labelled images.

Application of Deep Learning in Health Care

Artificial intelligence, machine learning and 'algorithmic learning' have been applied to medical data, including images from the earliest days of computing. Computer-assisted diagnostic systems have been around since the 1970s. Automated processing and analysis of one-dimensional time signals (e.g., electrocardiograms) has been around for decades. Computer-aided detection and diagnosis of medical images have also been around for quite some time, but gradually, the use of deep learning techniques is increasing in the health care sector. This is due to the following major reasons. First, the growth of machine-learning techniques, generally in many areas, especially in the use of unsupervised learning techniques, in latest application developments of Google, Facebook and IBM Watson. The second reason is the availability of healthcare data in public. The latter has effectively transformed medical records from carbon paper to silicon chips and made that data, structured and unstructured, easily accessible and available.

Mining Clinical Data for Faster Treatment

The medical records of a patient which may include prescription details, medical images, MRI scans, clinical test reports provide useful information to a doctor. Maintaining all these data of patients is a cumbersome job. Extracting useful information on time will help the doctor in diagnosis and detection of a disease. In these cases, GPU-accelerated DL tools can be used to process these data and find the relevant information of a patient for quicker treatment.

Quicker Diagnosis Process

Doctors rely greatly on medical images, like MRIs, CT-scans, X-Ray reports, Ultrasound reports, ECG signals to detect and diagnose diseases. At times these data are hard to analyse and consume a lot of time. In this case, DL can be used as an automatic tool to analyze the data, which can further help the medical practitioner to expedite the diagnosis process.

Prescribing Personalized Medicines

The genomic data are responsible for causing any disease in a living body. Identifying and studying the genomic data-causing disease is very difficult because genomic data are voluminous and extracting genetic parameters from huge genomic data are computation-intensive jobs and require quick processing. DL can be employed to handle voluminous genetic data for processing on these to mine information of disease-causing genetic materials, so that personal medicines can be prescribed to individual patients according to their genetic structure and analysis. A GPU-accelerated project has been taken over by a class of researchers from the University of Toronto working on a project named 'genetic interpretation engine'. Their work focuses on identifying cancer-causing genes present in individual patients.

Helping Disabled People

Visually-impaired people face a lot of difficulties while attending to their routine tasks, such as road crossing, recognizing people, reading tags and labels of different products. For this segment of people, wearable devices are being developed containing embedded prior-programmed chips. DL algorithms, with the help of other techniques like image processing, computer vision, are employed as embedded programs to perceive and understand the surroundings and transform these perceptions into words so that the blind people can learn about their surroundings and accomplish their tasks.

Prediction of Pharmacological Properties of Drugs

Experimental gene expression data collected from human cells are applied as input data to study the functional changes by using chemicals and drugs.

Table 1. Summary of areas and application in health care with input data.

Area	Input Data	Applications
Clinical Informatics	Digital clinical data Medical records	Disease prediction Human health monitoring Early notification on abnormality
Public Healthcare	Geospatial (Barik et al. 2016) images Social media Personal data captured from smart phone Text data	Air pollution detection Lifestyle disease controller Demographic information prediction Detection of infectious diseases epidemics
Medical Imaging	MRI/CITI scan reports X-Ray images Endoscopy pictures	Neural cells classification Healthy/unhealthy tissue classification Tumour detection Blood hemorrhage detection Cell clustering
Bioinformatics	DNA samples Gene sequence Microarray data	Drug designing Cancer diagnosis Gene classification Gene selection

The study focuses on the reaction of different newly-designed drugs by exploiting them to gene expression data. Such data are voluminous and hard to compute with conventional methods. Deep neural networks are used in these cases to forecast the properties of drugs on various diseases (Mamoshina et al. 2016).

Software Used in Deep Learning

Caffe

This is a deep learning framework which is developed by Berkly AI Research that facilitates a complete package for training, testing and adjusting the deep learning model parameters (CenterBerkeley 2016). It supports modularity so that the extension over the new version would be easy. It has a clear isolation between representation and execution layer. So switching between CPU and GPU is very easy and it just needs to call a function. Python and Matlab languages can be employed to create networks and for data classification, thereby supporting interface to bind these. It provides pre-trained reference models which help new researchers to quick start with work. In Caffe through configuration file, network architectural parameters are set and the network can be trained and tested through command line. This helps users not to write detail coding to configure and deploy the model. This tool is widely used for image recognition and computer vision task. It is applicable in medical image analysis. Some limitations of Caffe are that it is not well designed for recurrent network and customization can be made with C++ only (Jia et al. 2014).

Theano

Theano is built on a set of python library, is a declarative programming model aimed to provide research support in deep learning. Its native code is C++ and Python and innovated by University of Montreal. The core part of machine-learning application's execution is doing complex matrix operations. It provides a rich library to solve this mathematical expression with the help an intelligent compiler which uses heuristics to optimize the expressions and swiftly handle multidimensional arrays. It also has the support of GPU devices which run the expressions faster and with a higher degree of accuracy. It can be customized according to user application need. One drawback of Thiano is that the compilation time is more while dealing with larger datasets because the network model is produced as a computational graph and reconverted to native code through the optimized compiler (Universite de Montreal 2016).

Pylearn2

It is a machine learning library built in LISA at the University de Montreal. Since this is created for researchers, it has its own target users. The users must have expertise on machine learning algorithms to understand the code. It can execute symbolic expressions directly. It is built on top of Theano. It is aimed to be utilized for academic and scientific purposes. But it is a customized module that can be applied for any real data mining task, like classification, prediction, etc. The only limitation is that customization modules must be written in C++ (Goodfellow et al. 2013).

Caffe on Spark

It integrates the features of DL framework Caffe and big data framework apache spark and apache hadoop. It is an open-source framework that can create a distributed DL network with a cluster of CPU and GPU servers (Andy Feng, Jun Shi and Mridul Jain). Unlike some other DL framework which needs an isolated cluster for deep learning which requires large datasets to be migrated among cluster nodes for increased system latency and latency, here the cluster is not separate but is embedded in Spark application programme, by which unwanted data migration among cluster nodes is prevented (Kim et al. 2016).

NVIDIA cuDNN[]

cuDNN is a collection of primitives to implement standard routines of a deep neural network to create, train and evaluate. It is a GPU-accelerated framework which supports software applications build on created neural network models. It contains a number of tensor transformation functions which are helpful in repetitive computation done in ML algorithms. It can be integrated with Caffe and Baidu (Chetlur et al. 2014).

Keras

It is a Python library to implement deep learning which makes use of either Thiano or TensorFlow as backend. It runs in python 2.7 or 3.5 versions and can be used to design deep learning models, both for research and development. It is an open-source tool. The main guiding principle on which it is built is modularity and extensibility. Its native is Python and no other language is supported. It is used as a binary and multiclass classifier. It can also be used for regression analysis (Franois Chollet 2016).

Deep Learning 4j

Deep learning 4j is Java-based open-source library for creating and deploying deep neural network. It works in the company of Spark and Hadoop. Its native is Java having Apache 2.0 licence, so all JVM languages, such as Scala or Closurie, are compatible with it. It is written in C++ and can be integrated with Hadoop and Spark to execute on various other backends. It gives a plug-and-play solution instead of building configuration. As a part of deep learning 4j, ND4j is a vectorization tool containing a rich matrix library to do core machine learning tasks. This tool can be used to create recurrent network and supports reinforcement learning. Unlike other tools, it is solely built for commercial use in business houses. This framework is distributed and supports multi threaded programming model, which is capable to run in a parallel environment. It is a complete package to provide all sorts of applications to model a deep neural network possessing smart n-dimensional array class (Team 2016).

Mocha.jil

It is a DL library developed at MIT for a new programming language, Julia, especially designed to do scientific and numerical computation. Its native code is Julia and can interface with Julia only. It can work with Julia backend but supports GPU backend and can be integrated with Nvidia libraries, like cuDNN. It is highly modular and can be easily extended. It does not require configuration file for describing network architecture, which is directly written by Julia code. It contains some pre-trained models to do a quick start. The only difficulty with this is that it is compatible with Julia only (Internals 2015).

Torch7

Torch7 (Internals 2015, Collobert et al. 2016) was first built in the University of New York. Its foundation is on a special scripting language, called LuaJIT. Lua is chosen because it can easily integrate with C language. By placing appropriately the existing modules, deep networks can be formulated. As based on a scripting language, the model is not required to be compiled, thus making it suitable for quick prototyping if compared to Theano (Collobert et al. 2016). This is used in facebook AI lab and Google deepmind. A possible downside of this tool is that the user has to be familiar with the LuaJIT scripting language. Also, LuaJIT is not good at creating custom recurrent networks.

TensorFlow

Like Theano, TensorFlow is also an open-source library. It is designed to do numerical computation by using data flow graphs as in the case of neural network. It was innovated by the researchers of Google Brain Team at Google's Machine Intelligence laboratory. Unlike Thiano, TensorFlow is based on *distributed computing*, so is fit to work with many GPUs and CPUs in clusters. It has an inherent support for heterogeneous and parallel computation (Collobert et al. 2011). It is developed in C++ and has Python compatibility. One demerit of TensorFlow is that the pre-trained model is not available in a large proportion. It is used in handling biological data in bioinformatics to solve different problems like gene selection, prediction, etc. (Google 2016).

TensorGraph

It is an easier and simpler framework for creating a graphical model based on TensorFlow. It contains three types of nodes: StartNode which takes input for the graph, HiddenNode for creating the hidden layers and EndNode used for generating the outputs of the model (Rampasek and Goldenberg 2016).

Tinny-dnn

It is a header file written in C++ and contains some deep learning library. It can be easily integrated with any compatible application to build a network, which can be trained and tested by using suitable methods. It contains several activation functions such as tanh, sigmoid, softmax, leaky relu, etc. (Angelova et al. 2015). The broad division of these tools are depicted in Fig. 9. A comparative analysis on the available open source software is presented in Table 2. It gives a analogy on the software, their programming paradigm, suitability of application and their benefits and limitations.

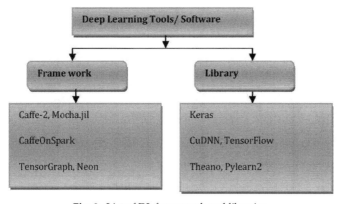

Fig. 9. List of DL framework and libraries.

Table 2. DL software with their pros & cons.

Software	Programming Paradigm	Suitable For	Pros & Cons
Caffe	Declarative	Image recognition Computer vision Speech recognition	*Pros* • User does not have to write detail code • Pre trained models are present to do a quick start *Cons* • Cannot be used to create recurrent network • Customization can be made only through C++
Theano	Declarative	Numerical computation	*Pros* • Good for tensor-based mathematical expressions • Compiler can optimize expressions *Cons* • Compilation time is more for larger datasets
Pylearn2	Declarative	Classification Prediction	*Pros* • Quicker computation *Cons* • Customization can be made only through C++ • User should have good knowledge on DL to understand the code execution
Keras	Emparative	Prediction Image classification Large-scale image recognition Feature extraction	*Pros* • Easily extendable *Cons* • Only interface is Python
Deep learning 4j	Object oriented programming	Object recognition in social networks for photo tagging	*Pros* • Can import neural net models from Tensorflow, Caffe, Torch and Theano • It is distributed • Complete package suitable for commercial use

Table 2 contd. ...

...Table 2 contd.

Software	Programming Paradigm	Suitable For	Pros & Cons
Torch7	Emparative	Used in Facebook AI research IBM Reaserch lab	*Pros* • Easily integrated with C • No compilation is needed *Cons* • User has to be familiar with LuaJIT
TensorFlow	Emparative	Image classification Drug Designing Information retrieval	*Pros* • Distributed and inherently parallel *Cons* • No pre-trained libraries
Neon	Object oriented	Image captioning Sentiment analysis	*Pros* • Optimization at assembler and data loading level • Rich graphical and visualization properties

Conclusion

In this chapter, the latest developments on deep neural networks are presented. Some widely-used deep learning architecture are investigated and selected applications for health care are depicted. Four classes of deep learning architectures, which are Restricted Boltzmann Machine, the Deep Belief Networks, the AutoEncoder, and the Convolutional Neural Network, are briefly discussed. Some of the current industry trends are also tried to be incorporated in this study. Different open-source tools used in DL are investigated and some of their details are provided. Nvidia cited that use of ML algorithm in health will change the dimension of current health care system. Earlier data were limited in size. So for mining information from data and using the same results to let a machine to estimate and predict something can be established by using simple neural networks having different learning rules. With the explosive increase in data, its dimension and structural composition are changing. As complexity and nonlinearity have been changed, conventional ML are not able to compute these data. DL came into picture a few years ago. Their architecture of having more intermediate computing layers is designed to handle the complexities and nonlinearity of big data. Some widely used tools like Caffe, Thiano,

Deep Learning 4j, Tensor Flow are discussed in the study by taking into consideration some of their technical specifications, their strengths and weaknesses. The health-related application areas are discussed along with specific problems, which are using DL tools and techniques. Most of the tools are designed for scientific and academic use. Full-fledged commercial tools are yet to come. Majority of the available tools are open source and can be customized according to one's requirement. As the technique is in its early stage, some concrete hypothetical theories are yet to be evolved.

Future Scope

Based on the study of literatures, some related issues for future research are presented below:

- **Unavailability of deep models for small sized training data:** DL has been found to be used with mainly in the domain of big data and has proved to be successful in the sard domain. However, on the contrary for limited-sized available data, the DL models are not being used. So there is need to investigate DL models for use in small-sized training data.
- **Uses of optimization algorithms to adjust the network parameters:** For finetuning the network parameters, several heuristic and meta-heuristic algorithms can be applied to adjust different parameters of the network, mainly in case of AE and DBN.
- **Unavailability of large training data:** For training deep neural network, a large amount of training data is required. This is causing a major problem in using DL with the present health care system, because the electronic data keeping system is yet to be achieved in this sector.
- **The visualization of DL technique:** It is still remains a black box for many researchers.

References

Angelova, A., A. Krizhevsky, V. Vanhoucke, A. S. Ogale and D. Ferguson. (2015, September). Real-time pedestrian detection with deep network cascades. *In*: BMVC (Vol. 2, p. 4).
Arel, I., D. C. Rose and T. P. Karnowski. (2010). Deep machine learning-a new frontier in artificial intelligence research [research frontier]. IEEE Computational Intelligence Magazine 5(4): 13–18.
Barik, R. K., H. Dubey, A. B. Samaddar, R. D. Gupta and P. K. Ray. (2016). FogGIS: Fog computing for geospatial big data analytics. pp. 613–618. *In*: Electrical, Computer and Electronics Engineering (UPCON), 2016 IEEE Uttar Pradesh Section International Conference on. IEEE.
Bengio, Y., P. Simard and P. Frasconi. (1994). Learning long-term dependencies with gradient descent is difficult. IEEE Trans. Neural Netw., 5(2): 157–166.

Bengio, Y. (2009). Learning deep architectures for AI. Foundations and Trends® in Machine Learning 2(1): 1–127.

CenterBerkeley, 'Caffe', (2016). [Online]. Available: http://caffe.berkeley vision.org/.

Chetlur, S., C. Woolley, P. Vandermersch, J. Cohen, J. Tran, B. Catanzaro and E. Shelhamer. (2014). Cudnn: Efficient primitives for deep learning. arXiv preprint arXiv: 1410.0759.

Collobert, R. and J. Weston. (2008, July). A unified architecture for natural language processing: Deep neural networks with multitask learning. pp. 160–167. In Proceedings of the 25th International Conference on Machine Learning. ACM.

Collobert, R., J. Weston, L. Bottou, M. Karlen, K. Kavukcuoglu and P. Kuksa. (2011). Natural language processing (almost) from scratch. Journal of Machine Learning Research, 12(Aug), 2493–2537.

Collobert, R., K. Kavukcuoglu and C. Farabet, "Torch," (2016). [Online]. Available: http:// http://torch.ch/.

Deng, L. (2012). Three classes of deep learning architectures and their applications: a tutorial survey. APSIPA Transactions on Signal and Information Processing.

Deng, L. (2012). Three classes of deep learning architectures and their applications: a tutorial survey. APSIPA Transactions on Signal and Information Processing.

Franois Chollet, "Keras," (2016). [Online]. Available: https://keras.io/.

Goodfellow, I. J., D. Warde-Farley, P. Lamblin, V. Dumoulin, M. Mirza, R. Pascanu and Y. Bengio. (2013). Pylearn2: a machine learning research library. arXiv preprint arXiv: 1308.4214.

Goodfellow, I., Y. Bengio and A. Courville. (2016). Deep Learning. MIT Press.

Google, "Tensorflow," (2016). [Online]. Available: https://www. tensorflow.org.

He, Kaiming, Xiangyu Zhang, Shaoqing Ren and Jian Sun. (2015). Delving deep into rectifiers: Surpassing human-level performance on imagenet classification. pp. 1026–1034. *In*: Proceedings of the IEEE International Conference on Computer Vision.

Hinton, G. E., S. Osindero and Y. W. Teh. (2006). A fast learning algorithm for deep belief nets. Neural Computation 18(7): 1527–1554.

Hinton, G. E. and R. R. Salakhutdinov. (2006). Reducing the dimensionality of data with neural networks. Science 313(5786): 504–507.

http://nptel.ac.in/courses/117106100/Module%208/Lecture%202/LECTURE%202.pdf.

http://www.deeplearning.net/tutorial accessed on 9-10-2017.

Hubel, D. H. and T. N. Wiesel. (1962). Receptive fields, binocular interaction and functional architecture in the cat's visual cortex. The Journal of Physiology 160(1): 106–154.

Internals, J. Proceedings of JuliaCon 2015 Research and Teaching Output of the MIT Community.

Jia, Y., E. Shelhamer, J. Donahue, S. Karayev, J. Long, R. Girshick and T. Darrell. (2014, November). Caffe: Convolutional architecture for fast feature embedding. pp. 675–678. *In*: Proceedings of the 22nd ACM International Conference on Multimedia. ACM.

Kim, H., J. Park, J. Jang and S. Yoon. (2016). DeepSpark: A Spark-Based Distributed Deep Learning Framework for Commodity Clusters. arXiv preprint arXiv: 1602.08191.

Krizhevsky, A., I. Sutskever and G. E. Hinton. (2012). Imagenet classification with deep convolutional neural networks. pp. 1097–1105. *In*: Advances in Neural Information Processing Systems.

LeCun, Y., Y. Bengio and G. Hinton. (2015). Deep learning. Nature 521(7553): 436–444.

Ling, Z. H., S. Y. Kang, H. Zen, A. Senior, M. Schuster, X. J. Qian and L. Deng. (2015). Deep learning for acoustic modeling in parametric speech generation: A systematic review of existing techniques and future trends. IEEE Signal Processing Magazine 32(3): 35–52.

Liu, W., Z. Wang, X. Liu, N. Zeng, Y. Liu and F. E. Alsaadi. (2017). A survey of deep neural network architectures and their applications. Neurocomputing 234: 11–26.

Mamoshina, P., A. Vieira, E. Putin and A. Zhavoronkov. (2016). Applications of deep learning in biomedicine. Molecular Pharmaceutics 13(5): 1445–1454.

Rampasek, L. and A. Goldenberg. (2016). Tensorflow: Biology's gateway to deep learning? Cell Systems 2(1): 12–14.

Team, D. J. D. (2016). Deep learning 4j: Open-source distributed deep learning for the JVM. Apache Software Foundation License, 2.

Universite de Montreal, "Theano," (2016). [Online]. Available: http://deeplearning.net/software/theano/.

Van den Oord, A., S. Dieleman and B. Schrauwen. (2013). Deep content-based music recommendation. pp. 2643–2651. *In*: Advances in Neural Information Processing Systems.

Williams, and D. Zipser (1989). A learning algorithm for continually running fully recurrent neural networks. Neural Comput., 1(2): 270–280.

Witten, I. H., E. Frank, M. A. Hall and J. Pal. (2016). Data Mining: Practical machine learning tools and techniques. Morgan Kaufmann.

Yu, D. and L. Deng. (2011). Deep learning and its applications to signal and information processing [exploratory dsp]. IEEE Signal Processing Magazine 28(1): 145–154.

CHAPTER **5**

Assessing the Effectiveness of IPTEACES e-Learning Framework in Higher Education
An Australian's Perspective

Tomayess Issa,[1,*] *Nuno Pena*[2] and *Pedro Isaias*[3]

Introduction

IPTEACES (acronym for Involvement, Preparation, Transmission, Exemplification, Application, Connection, Evaluation and Simulation) is an E-Learning Framework (Pena and Isaias 2010a, b, 2012, 2013). This was primarily developed through a pedagogical benchmark [mainly Gágne's Nine Events of Instruction (Gagné et al. 1992), Merrill's Principles of Learning (Merrill 2002, 2007), Keller's ARCS's model (Keller 2008) and van Merrienboer's et al., Ten Steps to Complex Learning (2007)] and by investigating award-winning e-courses (e.g., Brandon Hall Excellence in Learning Awards, International eLearning Association Awards) and corporate E-Learning best practices (e.g., Bersin & Associates reports). IPTEACES (Pena and Isaias 2010a, 2010b, 2012, 2013) addresses and

---
[1] Curtin University Perth Australia.
[2] ISEG - Lisbon School of Economics & Management, Portugal.
[3] The University of Queensland, Australia.
Emails: nuno.raposo.pena@gmail.com; pedroisaias@gmail.com
* Corresponding author: Tomayess.Issa@cbs.curtin.edu.au

assorted social-demography and geographically dispersed variety of attendees. It was originally conceived and designed as an instructional design framework for online corporate learning that could enable an integrated implementation of an appropriate learning strategy for different learners, considering different learning preferences and other specific differences (i.e., age, gender, educational background, previous knowledge in the area, literacy, computer proficiency, organizational culture, motivation, values and experience/inexperience in e-Learning) (Fig. 1). This study reports the application of IPTEACES Framework in a higher education institution in Australia, in two postgraduate units, in order to improve and enhance students' learning. This experience was conducted over a period of twenty-one months which included preparation, development, execution and examination of results. This study is organized as follows: (1) Introduction; (2) Theoretical Background; (3) Methodology and Research Question; (4) Participants and Assessment Tasks; (5) Results; (6) Discussion and Theoretical and Practical Significance; (7) Limitations; (8) Conclusion and Further Research Directions.

Theoretical Background

Previous work (Pena and Isaias 2010a, 2010b, 2012, 2013) details a description of each of IPTEACES learning phases (see Fig. 1):

- **Involvement:** This strategy aims to engage the student in the context of a real situation, where he is confronted with a problem (Merrill 2002, 2007). From a pedagogical perspective, it seeks to involve the student [Gagné's first event; Keller's (2008) first principle of ARCS].

- **Preparation:** This strategy is divided into two complementary stages: Presentation of 'Program and Objectives' and 'Contextualization and Activation' (a) Program and Objectives—Presentation of the program, objectives and what is expected of the student (Cf. Gagné's second event; Keller's second principle). (b) Contextualization and Activation strategy to make an introduction, a contextualization or a reminder of the subject so that the student can activate prior existing knowledge (Cf. Gagné's third event; Merrill's Activation principle).

- **Transmission:** This phase is divided into three complementary steps: acquisition (learning content), systematization and formative assessment. Acquisition is the central strategy for presenting the learning content of the course. This strategy (Gagné's fourth event) is where the new content is actually presented to the learner. After presenting a part of the new material, it is advisable to carry out systematization. It is also desirable, at the end, to create a graphical representation of the relationship between the concepts and ideas (new learning material) through the use, for instance, of 'concept maps' or 'dynamic diagrams.'

- **Exemplification and Demonstration:** This phase is mainly based on Merrill's (2002, 2007) 'Demonstration principle' and is divided into three complementary sub-strategies: Real Case, Step-by-Step Demo and Ask the Expert. (a) Real Case is an exemplification based on real cases and real situations and presents learners with authentic real-life situations, while illustrating the relevance of the content and demonstrating the concepts learned. (b) Step-by-Step Demo is a type of guided exemplification (Cf. Gagné's fifth event) that dissects a problem into phases and components and demands a detailed analysis of and commentary on the parts that constitute a complex situation or problem. (c) Ask the Expert phase presents the student with a more complex situation, a structured example, which may require the student to ask advice from an expert on how the problem can be resolved.

- **Application and Transfer:** This phase is an effort to maximize the transfer of learning by requiring students to flexibly apply what has been learned in new or unfamiliar situations (Cf. Gagné's fifth and sixth event; Keller's third principle and Merrill's application principle).

- **Connection:** This phase focuses on mentoring, collaboration and tools. (a) Asynchronous Mentoring - We developed for the course an integrated e-mail functionality enabling students to question their tutor. (b) Collaboration: Two kinds of discussion forums are available: supervised discussion forums and peer discussion forums) tools and which give the student access to a glossary of terms, job aids, documentation, worksheets, etc.

- **Evaluation Self-assessment and Summative Evaluation:** At the end of each learning module, the system suggests that the student submit a self-assessment. The intention is to determine whether the student, in his/her opinion, has achieved the learning objectives. Upon completing the modules, students are required to undertake a final assessment. This test, a summative evaluation, is intended to assess objectively whether the student has achieved the specific objectives of each of the learning modules. A detailed feedback follows the results of the summative assessment. This strategy relates directly to Gagné's eighth event and to Keller's fourth principle.

- **Simulation:** An exam was devised similar to the one that the candidates need to pass in the face-to-face examination after

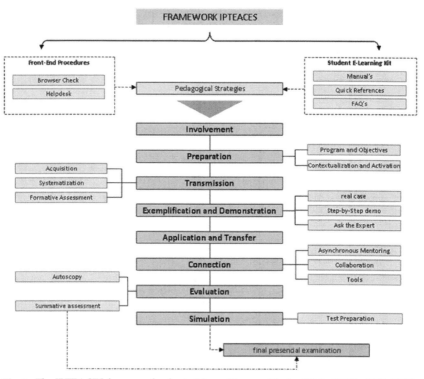

Fig. 1. The IPTEACES framework adapted from (Pena and Isaias 2010a, 2010b, 2012, 2013).

successfully completing all the e-Learning modules. This strategy takes into account Gagne's ninth event and Merrill's Integration Principle in particular.

Methodology and Research Question

This study will examine the question 'How can the IPTEACES e-Learning Framework satisfy students' learning in Australian tertiary education institutions?' To address this question, a mixed method approach was used in this research for gathering data, namely formal feedback, informal feedback, and an online survey designed and approved by the university ethics committee, and intended to examine students' attitudes to IPTEACES e-Learning Framework principles. Qualitative and quantitative techniques were utilized to collect the vital data to address the research question. The formal feedback collected from BPM, ITS units in Australia for 2013–2014 (semester 1) period, with the response rate being 80 per cent,

60 per cent, 52 per cent; 69 per cent; and 89 per cent respectively. This study addresses: 'How does the IPTEACES e-Learning Framework facilitate Lecturer's approach/work?' To obtain a comprehensive response, the lecturer's perspective will be included based on the preparation notes and assessment of the new learning module before and after the adoption, as well as students' reaction and feedback after using the new learning module.

Several studies (Harrison and Reilly 2011, Molina-Azorin 2011, Teddlie and Tashakkori 2009, Wiggins 2011) indicate that a mixture of research methods will assist research to address and answer the research questions and objectives in adequate depth and scope. The use of qualitative and quantitative techniques will minimize the chance of inconsistencies between the outcomes, thereby improving their strengths and decreasing their weaknesses. Mixed methods allow researchers to obtain a comprehensive understanding of students' reactions to the IPTEACES e-Learning Framework principles. It is expected that students' comments will assist researchers to enrich their teaching in future. Three sources are used in this study namely: formal feedback, informal feedback and an online survey. The formal feedback is a method used by the university to collect students' feedback about learning outcomes, experiences, resources, assessments, feedback, workload and overall satisfaction. The informal feedback is intended to assist the lecturer to modify and improve the unit before the end of the semester by fine-tuning certain aspects of the course in order to meet students' requirements and expand the methods of teaching and learning. The third source is an anonymous online survey which was distributed to students at the end of the semester. The survey was divided into three parts: Part 1: the demographic information; Part 2: level of satisfaction for the technology and support of online learning, online course content and leaner self-assessment when learning online; Part 3: level of importance for the technology and support of online learning, online course content and learner self-assessment when learning online. The online survey was adopted from Levy (2006) and Levy et al. (2008). They identified a set of characteristics built primarily from a review of literature, exploratory focus groups as well as in a qualitative questionnaire, that learners found important, or valuable, when using e-Learning systems. They developed an assessment of such characteristics using a survey instrument which included *satisfaction* and *value* items for each e-Learning system's characteristic, learners' overall value measure, overall satisfaction measure and an overall perceived learning measure.

In order to determine the level of effectiveness of this e-Learning project, Levy's methodology was applied. However, due to the specificity of this new learning module in the context of a Blended-Learning Course, we decided to withdraw some of the original items from the questionnaire that weren't directly related to the specificity of the project. Based on the output of the questionnaire, Levy (2006) and Levy et al. (2008) proposed two benchmark tools that should be complemented: 'The Value-Satisfaction grid' and 'LeVIS index' (see Fig. 2). The Value-Satisfaction grid is based on aggregated student perceived satisfaction as well as aggregated student-perceived value. This grid positions mean characteristics of satisfaction scores on the horizontal axis and the mean characteristics value scores are positioned on the vertical axis.

Despite the information given by the 'Value-satisfaction grid', this tool does not provide a measure of the effectiveness of an e-Learning system. Therefore, Levy proposed the 'LeVIS index' (Levy 2006, Levy et al. 2008) that provides a measure, an overall index of learners' perceived effectiveness, combining e-Learning systems value measures and e-Learning systems satisfaction measures. The 'LeVIS index' was proposed as the multiplication of the overall satisfaction (S∘) by the overall value (V∘). 'LeVIS index' provides a score of the overall magnitude of the effectiveness of the e-Learning system under study. The two items (S∘ and V∘) are measured on a scale of 1 to 6, and the 'LeVIS index' is calculated as:

The magnitude of LeVIS provides that when LeVIS is near 0, it indicates very low learners' perception of the effectiveness of e-Learning systems. When LeVIS is near 1, this indicates a very high learners' perceptions of e-Learning systems' effectiveness. This measure provides

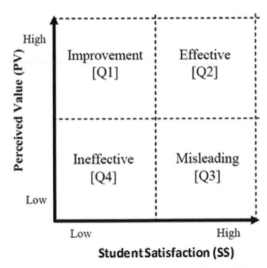

Fig. 2. The Value-satisfaction grid (adapted from Levy (2006, 2009)) prepared by the researchers.

that if only one of the two measures ($S_\circ$ or $V_\circ$) is high, the overall system measure (LeVIS) score is not high.

Participants and Assessment Tasks

This study involved students enrolled in three postgraduate units in Australia; the units are Business Project Management (BPM), Information Systems (IS) and Information Technology Seminar (Beavers et al. 2013) in Australia. The 262 participants were mainly from Australia, Asia (including India), Middle East, Europe, South America and Africa. Table 1 provides several demographic details of the BPM, IS and ITS students for the 2013–2014 period.

The units' assessment targets, skills and syllabus, which are designed mainly with university graduate attributes in mind (see Table 2).

Table 1. Postgraduate Participants—Australia (2013–2014)—prepared by researchers.

Unit	Students #	Gender		Countries					
		Female	Male	Australia	Asia (including India)	Middle East	Europe	South America	Africa
BPM	227	61	166	48	164	10	3	1	1
IS	13	1	12	0	9	4	0	0	0
ITS	22	7	15	4	15	1	1	1	0
Total	262	69	193	52	188	15	4	2	1

Results

The aim of this study was to examine students' attitudes to the new learning module (via IPTEACES e-Learning Framework). To conduct the research study, the researchers used three methods, namely: informal feedback, formal feedback and an online survey.

Students Informal Feedback

The informal feedback was distributed during the fourth week of the semester to obtain students' anonymous feedback to the teaching style and the assessment tasks of the new learning module (using IPTEACES e-Learning Framework) and Wiki. The majority of students confirmed that the new learning module assisted them to understand the unit materials and improve their skills. Students provided the following comments:

Table 2. Assessment tasks activities for postgraduate units—Australia (2013–2014)— prepared by researchers.

Unit	Assessment Tasks/Plus the Mark	This Assessment is Targets the Following Skills	Mark	Unit Syllabus
BPM	Individual assessment	Research, writing, reading, technology, information, critical thinking, endnote software.	20%	BPM provides a framework for project management, its context and processes. Students study a range of topic areas, including management of scope, time, cost, quality, communications and risk, as applied to business projects; team development; cost estimation techniques and tools; risk management, and project management standards and accreditation. Students will also learn and utilize contemporary project management software packages in the unit.
	Team work assessment	Teamwork, research, writing, reading, technology, information, critical thinking, endnote software.	30%	
	Final exam	Critical thinking and decision making.	50%	
IS	Mini tests	Critical thinking and decision making.	25%	IS unit is primarily centered on usability and human-computer interaction (HCI) principal's design toward users' interface, including website.
	Reflective journals (7 individual and 3 teamwork)	Research, writing, technology, information, critical thinking, reading, and endnote software.	40%	
	Contribution to group discussions—Blackboard	Writing, reading, debate, written presentation and oral skills and drawing (i.e., concept maps).	10%	
ITS	Three journals	Research, writing, technology, information, critical thinking, reading, and endnote.	30%	The ITS unit focuses mainly on issues relating to strategic development, IT business, sustainability tools and Green IT and other related issues.
	Individual presentation of an IT sustainable strategy and report writing	Digital oral presentation; reading, writing, research, search, critical thinking, and endnote software.	55%	
	Wiki for collaborative writing	Writing, reading, debate, written presentation and oral skills and drawing (i.e., concept maps).	15%	

- *As postgraduate course, I expected a dull and sober class environment which I realized was not true when the classes started as the subject itself had been framed interactively using a website, and I must say that the developers working behind the website continue to do a great job, as the course material insists on interactive user participation by including interesting quizzes in between the topics. It is evident that the developers of academia have definitely stuck to the HCI [Human Computer Interaction] guidelines and principles!*
- *The material (new learning module via IPTEACES) gave me a better understanding of this unit.*
- *Very interesting slides (new learning module via IPTEACES) and hopes to have more.*
- *Presentation slides (new learning module via IPTEACES) were useful and enjoyable. It was a healthy and useful experience. I would be glad if this is kept going and fresh.*

Students' feedback indicates that the new learning module enriched their learning skills by enabling them to login to the website from any device at any convenient time and place. From the informal feedback, a meta learning approach was applied in this study to serve as guidelines and principles of Human Computer Interaction in the new learning process, as students acknowledged the Portuguese team endeavour. Several studies (Munoz-Organero et al. 2010, Rennie and Morrison 2013, Sridharan et al. 2010) confirmed that technology use in teaching and learning motivates students to learn the unit materials in an efficient way, since the new learning modules are formative and creative. Students appreciated the mixed teaching approaches as these allowed them to enhance their knowledge, personal and professional skills for this study as well as for the future.

Students Formal Feedback

The students' feedback is collected by the university at the end of each semester. This feedback is given anonymously and provides qualitative and quantitative information about the unit. Table 3 shows the quantitative results for the unit resources as well as the overall satisfaction in 2013–2014 which ranging from 84–100 per cent. These results confirmed that students are pleased with the unit resources, the lecture notes, and the new learning module. Results presented in Table 3 confirm that the current resources, assessments and activities in BPM, IS, and ITS motivate students to acquire the necessary knowledge from these units and enhance their personal and professional skills as they are strongly motivated to participate in all activities and complete assessment tasks. The formal feedback produced the following qualitative comments from students:

Table 3. eVALUate results for period 2013–2014—prepared by researchers.

Year/Semester/ Unit	Students Enrolled	Responses Rate	Item 3: The Learning Resources in this Unit help me to Achieve the Learning Outcomes		Item 11: Overall, I am Satisfied with this Unit	
			Unit	University Average	Unit	University Average
BPM/2013/S1	40	80%	100%	85%	97%	84%
BPM/2013/S2	42	60%	92%	85%	84%	84%
BPM/2014/S1	89	52%	96%	86%	98%	84%
IS/2013/S1	13	69%	100%	85%	100%	84%
ITS/2013/S2	13	69%	100%	85%	100%	84%
ITS/2014/S1	9	89%	100%	86%	100%	84%

- *They [new learning module via IPTEACES], are very nice and clear; the images are too good and they attract all the attention.*
- *Very interesting, interactive presentations [new learning module via IPTEACES], I learned a team from Portugal created and developed this new work; good job and well done.*
- *Presenting information [new learning module via IPTEACES] about this tool is interesting and liked it.*
- *The new notes' structure [new learning module via IPTEACES] is helping students to understand the unit better and in an easy way.*
- *The learning experience from the unit as well presentation [new learning module via IPTEACES] was so enriched and enjoyable that I cannot come up with any recommendations. It was a healthy and useful experience. I would be glad if this is kept going and fresh.*
- *Extra effort from lecturer.*

Several studies (Ghazizadeh et al. 2012, Issa 2014, Issa et al. 2012, Kuzma 2011, Mahruf et al. 2012, Mi and Gould 2014) show that technology use in teaching will encourage students to develop independent learning skills and be more engaged in the learning process. Table 3 shows that BPM, IS, and ITS students are pleased with unit resources [item 3], as this facility helped them achieve the unit learning outcomes and improve their professional and personal skills. Furthermore, item 11 'Overall

Satisfaction' confirmed that lecturer's teaching approach played a major role in increasing the students' commitment and satisfaction with the unit.

Students Feedback — Online Survey

The participants for this study are members of the Internet Generation from Australia and countries like: Asia (including India), Middle East, Europe, South America and Africa (see Table 1). Table 4 shows the gender component: 79 per cent male and 21 per cent female, as well as the response rate for the online survey.

Regarding the education level of participants, it was noted that the majority of Australian respondents had a Bachelors degree and Masters degree (see Table 5).

As for the online survey response rate, 91 per cent are from BPM, while the rest from ITS unit completed the whole survey (see Table 6). The valid response rate is 53.8 per cent. The IS unit did not participate in the online survey; however, students generously provided outstanding feedback through informal and formal comments.

A Likert six-point scale (i.e., ranging from extremely unsatisfied to extremely satisfied) was used in the online survey, especially for sections two and three and the participants were given the opportunity to provide additional comments after each section (Likert 1932). All the pages of the survey contained instructions at the top of the page and a progress bar at the bottom to provide feedback to users about their proximity to completion. A formal letter with the survey link was e-mailed to the students. Pages 1–4 presented the survey items with three questions per page to minimize scrolling, and the concluding page thanked the respondents for their participation. A description of each part was provided to the participants to explain its purpose. Table 7 presents the 21 system characteristics of this study, as well as the scores of standard deviation, mean satisfaction, mean importance and the LeVIS score from the responses of the 143 students.

Standard Deviation is a measure of how spread out or dispersed the data in a set are relative to the set's mean. Table 7 shows that system characteristics 'A03 – Amount of material in courses' (1.045) and 'A12 – Taking practice tests prior to graded test' (1.033) had the highest Standard Deviation value. These values indicated and confirmed students' satisfaction towards, the new learning module via IPTEACES, as students read, reflected and practiced the notes, and gathered the essential knowledge for this unit. On the other hand, system characteristics 'B09

Table 4. Number and percentage of online survey and gender prepared by researchers.

Number and Percentage of Online Survey	
Questionnaires distributed	262
Questionnaires returned	143
Response rate	54.6%
Gender	
Male respondents	113 (79%)
Female respondents	30 (21%)

Table 5. Highest education level—prepared by researchers.

Highest Education Level	Response	%
Primary education	0	0%
Higher secondary/pre-university	6	6%
Professional certificate	3	3%
Diploma	2	2%
Advanced/higher/graduate diploma	1	1%
Bachelor's degree	44	42%
Postgraduate diploma	5	5%
Master's degree	45	42%
Total	106	100%

Table 6. Online survey unit response rate—prepared by researchers.

Unit	Response
BPM501	70 (91%)
ITS653	7 (9%)
Total	77 (100%)

– Access of all courses from one area (My Black Board)' (0.868) and, 'A11 – Organization of courses (content of courses, organization of assignments, etc. across all course modules)' (0.883) had the lowest Standard Deviation values, as students preferred to have all the material, including the new learning module via IPTEACES in one place, not moving from one location to another.

Table 7. e-Learning system characteristics (adopted from Levy 2006, Levy et al. 2008)—prepared by researchers.

	#	Question	SD	Mean SAT	Mean IMP	LeVIS Score
B04	1	Access to courses from anywhere in the world (via the Internet)	.970	4.73	5	0.66
B06	2	Learning at any time of the day (schedule flexibility)	.969	4.86	5	0.68
B07	3	Submit assignments from anywhere (via the Internet)	.899	4.85	5.04	0.68
B08	4	Different system tools (chat, bulletin-board or discussion forums, etc.)	.892	4.64	4.81	0.62
B09	5	Access of all courses from one area (My Black Board)	.868	4.77	5.12	0.68
B11	6	Review course materials	.928	4.67	5.08	0.66
A01	8	Availability of course content	.908	4.57	5.13	0.65
A02	9	Quality content of courses	.881	4.69	5.15	0.67
A03	10	Amount of material in courses	1.045	4.48	4.96	0.62
A04	11	Interesting subject matter	.993	4.55	4.92	0.62
A05	12	Difficulty of subject matter	.952	4.36	4.76	0.58
A06	13	Availability of other contents (syllabus, objectives, assignments, schedule)	.946	4.45	5	0.62
A07	14	Enjoyment from the courses/ lessons	.990	4.53	4.96	0.62
A08	15	Ease-of-use (with course content, navigation, interface, etc.)	.952	4.57	4.91	0.62
A09	16	Similar to interface across all online courses	.926	4.52	4.76	0.60
A10	17	Gathering information quickly	.975	4.53	4.98	0.63
A11	18	Organization of courses (content of courses, organization of assignments, etc. across all course modules)	.883	4.56	4.92	0.62
A12	19	Taking practice tests prior to graded test	1.033	4.36	4.88	0.59
A13	21	Learning a lot in these classes	.973	4.47	4.93	0.61

Value-satisfaction Grid and LeVIS Index

The following sections will analyse in detail the variables of Satisfaction, Importance and LeVIS Scores. Figure 3 shows that several e-Learning system characteristics were placed in the 'Q1 – Improvement' quadrant, more specifically 'A05 – Difficulty of subject matter' and 'A12 – Taking practice tests prior to graded test'. The other three, 'A13 – Learning a lot in these classes', 'A03 – Amount of material in courses' and 'A06 – Availability of other content (syllabus, objectives, assignments, schedule)' are positioned very near to 'Q2 – Effective' quadrant.

Given the results of LeVIS index of Dimension A - Content, presented by Fig. 4, the global score reveals a 'good effectiveness' (> 0.56 and < 0.75) having a response score average of 0.62. More specifically, and according to Fig. 4, it is possible to conclude that the highest scores belong to the e-Learning System characteristics 'A2 – Quality content of courses' (0.67) and 'A1 – Availability of course content' (0.65). On the other hand, characteristics 'A5 – Difficulty of subject matter' (0.58) and 'A12 – Taking practice tests prior to graded test' (0.59) are the ones with the lowest scores, indicating moderate effectiveness.

The 'Value-Satisfaction grid' for Dimension B – Asynchronous/ Value presented in Fig. 5, illustrates that all of the six e-Learning system characteristics were placed in the quadrant of high satisfaction and high value, i.e., in 'Q2 – Effective' quadrant. The highest Importance Score was attributed to 'B11 – Review course materials' (Importance 5.08) and

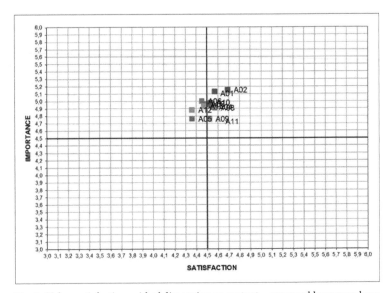

Fig. 3. Value-satisfaction grid of dimension a: content—prepared by researchers.

'B7 – Submit assignments from anywhere (via the Internet)' (5.04). The highest Satisfaction Score belongs to 'B6 – Learning at any time of the day (schedule flexibility)' (4.86) and 'B7 – Submit assignments from anywhere (via the Internet)' (4.85).

Figure 5 also shows that although placed in the Effective quadrant (Q2), items 'B9 – Access of all courses from one area (My Black Board)'and 'B11 – Review course materials' need to increase the level of satisfaction due to the fact that importance is viewed by users as being very high.

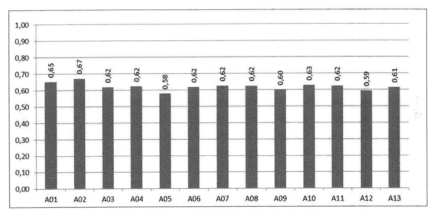

Fig. 4. LeVIS index of dimension a: Content—prepared by researchers.

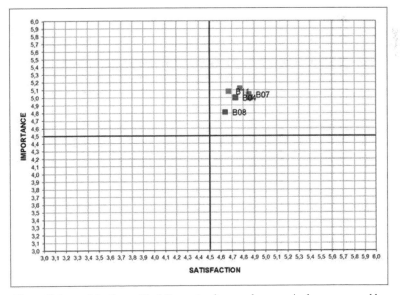

Fig. 5. Value-satisfaction grid of dimension b: asynchronous/value—prepared by researchers.

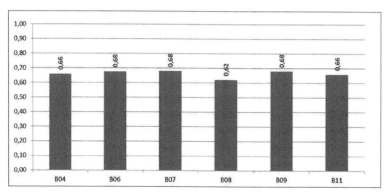

Fig. 6. LeVIS index of dimension b: asynchronous/value—prepared by researchers.

Given the results of LeVIS index of Dimension B – Asynchronous/Value, Fig. 6, the global score reveals that 'good effectiveness' (> 0.56 and < 0.75) had a response score average of 0.66. More specifically, according to Fig. 6, it is possible to conclude that the highest scores belong to 3 items, specifically, 'B06 – Learning at any time of the day (schedule flexibility)', 'B07 – Submit assignments from anywhere (via the Internet)' and 'B09 – Access of all courses from one area (My Black Board)', reaching a score of 0.68. On the other hand, the lowest scored characteristic was 'B08 – Different system tools (chat, bulletin-board or discussion forums, etc.)' (0.62).

Levy (2006) and Levy et al. (2008) proposed the following categorization for LeVIS index overall scores:

- If LeVIS overall score is ≥ 0.9375 – Very high effectiveness
- If LeVIS overall score is ≥ 0.75 and < 0.9375 – High effectiveness
- If LeVIS overall score is ≥ 0.5625 and < 0.75 – Good effectiveness
- If LeVIS overall score is ≥ 0.3750 and < 0.5625 – Moderate effectiveness
- If LeVIS overall score is ≥ 0.1875 and < 0.3750 – Low effectiveness
- If LeVIS overall score is < 0.1875 – Very low effectiveness.

As can be seen from Fig. 6, the results from the Global LeVIS index indicate that the overall e-Learning system (19 items) under study reached a global score of 0.63; therefore users classified it as having 'good effectiveness'.

Overall Value-Satisfaction Grid of Two Dimensions (All Items)

The collected results showed in Fig. 7 demonstrate a good overall effectiveness. None of the e-Learning system characteristics were placed

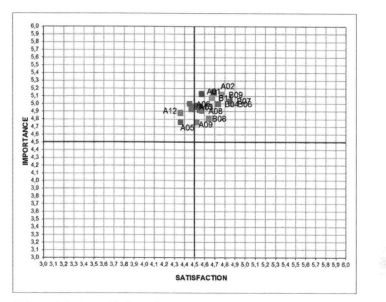

Fig. 7. Value-satisfaction grid of two dimensions (all 19 items)—prepared by researchers.

in 'Q4 – Ineffective' quadrant nor in the 'Q3 – Misleading' quadrant. The majority of items were concentrated in the quadrant of high satisfaction and high value, i.e., in 'Q2 – Effective' quadrant. However, some items from 'Dimension A – Content' showed moderate effectiveness and therefore should be the priority of quality improvement measures; more specifically 'A05 – Difficulty of subject matter' and 'A12 – Taking practice tests prior to graded test'. The other three, 'A13 – Learning a lot in these classes', 'A03 – Amount of material in courses' and 'A06 – Availability of other content (syllabus, objectives, assignments, schedule)' are positioned very near 'Q2 – Effective' quadrant and therefore should be subject in the short term to some improvement measures in order to maximize satisfaction (as they are considered important by users).

Lecturer's Perspective

From the lecturer's perspective, this experience is exciting and an outstanding achievement, as students indicated that the new learning module allowed them to obtain the necessary knowledge and improve their skills. The development and implementation of the new learning module is a challenging exercise as it requires preparation, development, testing, execution and maintenance. Figure 8 demonstrates the steps to carry out when implementing the IPTEACES e-Learning Framework in

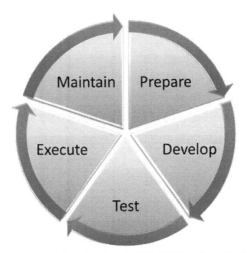

Fig. 8. The concept model for adopting the IPTEACES e-Learning Framework in Australia—prepared by researchers.

higher education. This section aims to discuss the lecturer's perspective in terms of these phases and reveal the lecturer's opinion of the strengths and weaknesses associated with the adoption of IPTEACES e-Learning Framework in postgraduate units.

The new learning module developed via IPTEACES e-Learning Framework required twenty-one months for the completion of all the concept model stages (see Fig. 8). The preparation stage was completed by the Australian lecturer based on the unit objectives and program; for BPM unit, two PowerPoint files were utilized: review one contains notes and questions from lectures 1 to 6; while review two is similarly structured and covers lectures 7–12. Each review consists of 41 slides. For the ITS unit, three PowerPoint files were developed, namely, Sustainability and Green IT (32 slides); Sustainability: what does this mean for business? (24 slides); Teece and Porter's Models (62 slides). For the IS unit, three PowerPoint files were developed, namely, HCI and Usability (48 slides); Evaluation, Design Principles and Guidelines (115 slides) and Usability and Interface (87 slides). Later, the Portuguese team developed the new learning module based on the preparation stage and added multiple-choices questions to ensure that the IPTEACES e-Learning Framework was implemented for each file.

The testing stage was carried out by lecturers in both the countries, to determine whether the new learning module was aligned with the needs of the Australian and Portuguese teams. To implement and apply security and privacy aspects of this study, the execution stage was implemented by sharing the surname and password for each student, since the new learning module was uploaded to the Portuguese's server so that it can be effectively sustained and maintained. Finally, the maintenance stage is essential in this model for further assistance if required, both now and in future. The Portuguese team was very helpful and cooperative during the whole process. An ethics approval from the Australian university was obtained to carry out this research.

The whole experience was intended to examine students' satisfaction with and perception of the importance of the new learning module style, as well as other aspects, such as the navigation, ease of use, and most importantly, whether this approach delivered the necessary knowledge and skills. The results confirmed that students are satisfied with the new learning module and knowledge acquired, as they perceived that their learning experience was greatly enhanced. Finally, the formal feedback confirmed that the learning resources in these units allowed students to achieve the learning outcomes, as the percentages ranged from 92–100 per cent, while the overall satisfaction ranged from 84–100 per cent. These percentages provide accurate quantitative evidence to support the research aims and have encouraged the Australian and Portuguese teams to continue to incorporate the latest e-Learning tools in their teaching approach.

From the lecturer's perspective, this study strongly supports the notion that the implementation of the new learning module makes units more interesting, motivating, exciting, attractive, interactive and appealing. Students obtain better understanding of the course content, and the whole process is useful, enjoyable and rewarding. However, one weakness, from the lecturer' perspective, is having to transpose the new learning module to power-point slides to allow the students to study them in the traditional way, instead of printing each slide. Moreover, saving mode is unavailable and inaccessible to lecturer and students; therefore, students need to be online continuously to acquire and review these notes. This exercise is a challenging task not only in the preparation stage, but in all the stages, as the Australian lecturer should examine whether the whole process is running smoothly from the perspective of students and the Portuguese lecturers. Finally, the lecturer's approach to teaching becomes unique and exceptional as compared to the way in which other units are delivered, as indicated by the following student comments:

- *During the learning process, students used many learning tools such as Wiki and presentation tools [new learning module via IPTEACES], which strongly stimulate the desire to learn. Hence, the unit should retain these teaching tools and encourage their adoption in other units.*

- *Overall satisfaction of learning online is that I can learn at anytime and anywhere. The availability of the course material as well as the other information and supports offer a convenient way of studying without a heavy book or a lot of printed papers.*

The whole experience was astounding and motivating for both the lecturer and the students. The students' feedback confirmed their satisfaction and acknowledgement of the lecturer's efforts in developing and presenting these notes. Finally, is this experience worthwhile? The answer is most definitely 'yes', since via the IPTEACES e-Learning Framework, students learn and acquire the necessary knowledge to do well. The application of the latest technological tools for the purposes of teaching and learning will make our classes unique when compared to other units.

Discussion, Theoretical and Practical Significance and Limitation

This study provides a strong basis for the adoption of the IPTEACES e-Learning Framework in tertiary education, in Australia. This experience required a total of twenty-one months for preparation, development, execution and examining the results obtained from 262 students. It was a challenging exercise for the Australian and Portuguese teams and students. The study results are encouraging, inspiring and reassuring since the IPTEACES e-Learning Framework contains the necessary stages for developing an outstanding e-Learning resource for students (Internet generation), since at present they rely heavily on ICT.

Besides providing the study questions, this new learning module satisfies the students' learning in an Australian tertiary institution, as the global score from the online survey was 0.63, which means that users classified it as having 'good effectiveness'. Furthermore, students' comments, derived from three types of data collection, confirmed this: *"I am very satisfied with the resources (new learning module via IPTEACES) and the ease of accessing the course-related material online,"* and *"mobility and ease of access are the most important elements for studying online"*.

The IPTEACES e-Learning Framework facilitates the lecturer's approach/work by incorporating the new learning module (via IPTEACES) in her teaching and students confirmed their satisfaction and eagerness to accept this approach. The Australian lecturer confirmed that her unit appeared to be unique in the university and students insisted that this teaching and learning method be introduced in other university courses since the module was informative and easy to use. The new learning module allowed the lecturer to meet students' needs as the online survey results confirmed the students' satisfaction and enjoyment. It assisted them to understand the aims of the unit, and their learning experience was enjoyable and quite unique. The online overall e-Learning system (19 items) under study reached a global score of 0.63, indicating that users classified it as having 'good effectiveness'. The research questions are answered and this study will assist other lecturers to implement this new learning module and meet students' needs and requirements.

The findings of this study have made significant contributions to the research and literature in the realm of information technology, particularly in terms of positive student outcomes related to students' knowledge, learning and skills. Moreover, the lecturer's unique approach has given her notable distinction within the university.

This study is limited to 262 students from Australia. The rationale behind this study is a collaborative research project undertaken by an Australian lecturer and two Portuguese lecturers to examine students' and lecturer's attitudes to a new learning module (via IPTEACES e-Learning Framework). Further research with larger and more diverse groups of students is required in the future to strengthen the research findings. Finally, the authors will take into consideration the students' feedback about locating all the notes, including the new learning module via IPTEACES at one location.

Conclusion and Further Research Directions

This study aims to examine students' and lecturer's attitudes to a new learning module (via IPTEACES e-Learning Framework) in three postgraduate units in Australia. This experience was a challenging exercise for both Australia and Portugal as it spanned a total of twenty-one months covering preparation, development, execution and examination. However, the results confirmed that the new learning module allowed students to understand the unit material and enhance their professional and personal skills, while the lecturer delivering the course was distinguished by his novel approach. Finally, the IPTEACES e-Learning Framework, showed, again and in a completely different environment, very positive results in terms of the effectiveness of the learning program and it would be very interesting to promote, apply and analyse IPTEACES e-Learning Framework in other contexts, academia and corporate settings, as well as in the adaptation to specific populations and subjects. Further research will be carried out in future in order to compare Brazil and Portugal to Australia.

References

Beavers, Amy S., John W. Lounsbury, Jennifer K. Richards, Schuyler W. Huck, Gary J. Skolits and Shelley L. Esquivel. (2013). Practical considerations for using exploratory factor analysis in educational research. Practical Assessment, Research & Evaluation 18(6): 1–13.

Gagné, R. M., L. J. Briggs and W. W. Wager. (1992). Principles of Instructional Design Edited By 4th. Forth Worth, Tx: Harcourt Brace Jovanovich College Publishers.

Ghazizadeh, M., J. D. Lee and Linda Ng Boyle. (2012). Extending the technology acceptance model to assess automation. Cogn. Tech Work 14: 39–49.

Harrison, R. and T. Reilly. (2011). Mixed methods designs in marketing research. Qualitative Market Research: An International Journal 14(1): 7–26.

Issa, Tomayess, Theodora Issa and Vanessa Chang. (2012). Technology and higher education: An Australian study. The International Journal of Learning 18: 223–236.

Issa, Tomayess. (2014). Learning, communication and interaction via wiki: An Australian perspective. pp. 1–17. *In*: Harleen Kaur and Xiaohui Tao (eds.). Icts and the Millennium Development Goals. Usa: Springer.

Keller, J. (2008). First principles of motivation to learn and e-Learning. Distance Education 29(2): 175–185.

Kuzma, J. (2011). Using online technology to enhance student presentation skills. *In*: Worcester Journal of Learning and Teaching. Worcester, Uk: University of Worcester.

Levy, Y. (2006). Assessing the Value of E-Learning Systems. Hershey, Pa: Information Science Publishing.

Levy, Y., K. E. Murphy and S. H. Zanakis. (2008). A value-satisfaction taxonomy of is effectiveness (Vstise): A case study of user satisfaction with is and user-perceived value of is. International Journal of Information Systems in The Service Sector 1(1): 93–118.

Likert, R. (1932). A technique for the measurement of attitudes. Archives of Psychology 22(140): 55.

Mahruf, M., C. Shohel and Adrian Kirkwood. (2012). Using technology for enhancing teaching and learning in Bangladesh: Challenges and consequences. Learning, Media and Technology 37(4): 414–428.

Merrill, M. David. (2002). First principles of instruction. Educational Technology Research and Development 50(3): 43–59.

Merrill, M. David. (2007). First principles of instruction: A synthesis. *In*: Reiser, R. A. and J. V. Dempsey (eds.). Trends and Issues in Instructional Design and Technology. Columbus: Ohio: Merrill Prentice Hall.

Mi, Misa and Douglas Gould. (2014). Wiki technology enhanced group project to promote active learning in a neuroscience course for first-year medical students: An exploratory study. Medical Reference Services Quarterly 33(2): 125–135.

Molina-Azorin, J. (2011). The use and added value of mixed methods in management research. Journal of Mixed Methods Research 5(1): 7–24.

Munoz-Organero, Mario, Pedro Munoz-Merino and Carlos Delgado Kloos. (2010). Personalized serevice-oriented e-Learning environments. IEEE Internet Computing 14(2): 62–67.

Pena, N. and P. Isaias. (2010a). An approach to diversity: The effectiveness of IPTEACES e-Learning framework. *In*: Proceedings of The 9th European Conference on elearning, Ecel. Porto-Portugal.

Pena, N. and P. Isaias. (2010b). The Ipteaces e-Learning framework—The analysis of success indicators and the impact on student social demographic characteristics. *In*: Proceedings of Iadis International Conference on Cognition and Exploratory Learning in Digital Age (Celda). Timisoara, Romania.

Pena, N. and P. Isaias. (2012). The IPTEACES e-Learning framework: Success indicators, the impact on student social demographic characteristics and the assessment of effectiveness. pp. 151–169. *In*: Pedro Isaias, Dirk Ifenthaler, Demetrios G. Sampson and

J. Michael Spector (eds.). Towards Learning and Instruction in Web 3.0. Springer New York.

Pena, N. and P. Isaias. (2013). Assessing the effectiveness of an e-Learning framework: The Portuguese insurance academy case. Journal of Cases on Information Technology 15(1): 1–18. Doi: 10.4018/Jcit.2013010101.

Rennie, Frank and Tara Morrison. (2013). E-Learning and Social Networking Handbook Resources for Higher Education. Edited By 2nd Edition: Routledge.

Sridharan, Bhavani, Hepu Deng, Joyce Kirk and Brian Corbitt. (2010). Structural equation modelling for evaluating the user perecptions of e-Learning effectiveness in higher education. *In*: 18 European Conference on Information Systems South Africa.

Teddlie, C. and A. Tashakkori. (2009). Foundations of Mixed Methods Research—Integrating Quantitative And Qualitative Approaches in The Social And Behavioral Sciences. Usa: Sage Publisher.

Van Merrienboer, J. J. G. and P. A. Kirschner. (2007). Ten steps to complex learning. Mahwah, Nj: Lawrence Erlbaum Associates.

Wiggins, B. (2011). Confronting the Dilemma of mixed methods. Journal of Theoretical and Philosophical Psychology 31(1): 44–60.

CHAPTER **6**

Review Study
CAPTCHA Mechanisms for Protecting Web Services against Automated Bots

Mohamed Torky[1,3] and *Aboul Ella Hassanien*[2,3,*]

Introduction

The rapid growth of internet and Cloud Computing technologies has made it is necessary for web servers to immunize and protect its online services and resources against automated tools called Robot (Bot) or scanner. Bot is defined as a computer program which is able to run a set of sequential operations continuously in automated fashion without any need of human interaction. Also, it is implemented as a single software system that retrieves information from remote sites using standard web protocols (Alessandro and Bergadano 2010, Alessan-dro and Sicco 2009). With increasing number and complexity of online services and resources without security of web applications, Bot become a risky threat to these services and resources by misusing its integrity, security or availability. For example, Robot software can perform a set of malicious activities for achieving penetrating purposes, such as content mirroring, e-mail harvesting, account/e-mail registration, Denial of Services (DoS), requesting fake communication, brute force attack and flooding web blogs.

Robot (Bot) technology passed over three generations (Alessandro and Sicco 2009) with the first generation included at automated tools able

[1] Faculty of Science, Menofyia University, Egypt; Email: mtorky86@gmail.com
[2] Faculty of Computers and Information, Cairo University, Egypt.
[3] Scientific Research Group in Egypt (SRGE); http://www.egyptscience.net
* Corresponding author: aboitcairo@gmail.com

to automatically retrieve a set of predefined resources, without seeking to interpret the content of downloaded resources. The second generation included automated tools able to analyze HTML pages through searching links to other resources and elaborate simple client-side code. In some cases, they can also record and repeat a series of user clicks and data insertion to mimic the human behaviors. The third generation includes smart tools which completely interpret client-side languages and understand meanings of web contents in a way more similar to what a human does. So, this generation is more intelligent than the other two generations (Proffitt 2013). Since the future of automated Robots is rapidly moving towards the development of more complex and smart programs, proper anti-Bot techniques are required to prevent the risky activities of such intelligent agents. The best approach to resist Bots is Completely Automated Public Turing Test to Tell Computer and Human Apart (CAPTCHA) (Kanika and Chadha 2013). Such an approach was formalized in 2000, by von Ahn et al., after being challenged by Yahoo's chief scientist to devise a security technique able to keep Rbots software out of Yahoo's web chat (Von Ahn et al. 2003). CAPTCHA is considered a challenge response technique to distinguish the human behaviors rather than Bot behaviors. The challenge is presented through an interface which asks the party that aims to access a specific resource or service to solve the CAPTCHA test to be an authorized party. CAPTCHA nature and design is based on the appreciated intelligence gap between human and machine ability with respect to some problem, which is easy to solve for humans but not yet solvable by computers. Hence, it is possible to develop effective mechanisms to tell them apart. In the last few years, several classes of CAPTCHAs have been proposed, based on hard-to-solve Artificial Intelligence (AI) challenges (Von Ahn et al. 2004). In this study, we review two generation (i.e., classical and modern) of CAPTCHA tests and discuss the more common types of CAPTCHA in each generation. In addition, we present the most risky attacks that aim to break CAPTCHA system automatically. Our review study ends with some key security principles to design a good and strong CAPTCHA test.

The rest of this chapter is organized into Six sections. Section 2, CAPTCHA Architecture, presents the architecture of CAPTCHA system. Section 3, Classical CAPTCHA Generation, discusses the known classical CAPTCHA approaches. Section 4, Next CAPTCHA Generation, Introduces new classes of CAPTCHA based on a novel designs which represent the next generation of CAPTCHA. Section 5, Attack Models and Breaking Techniques, discusses some of the common risky attacks that target CAPTCHA system . Section 6, Security Principles for Designing Secure CAPTCHA ,provides some of key security principles to design and develop any CAPTCHA system. Section 7, Conclusion, summarize the general conclusion of this study.

CAPTCHA Architecture

A Web Server such as Facebook Server may be holding both public and protected resources and services that may be in the form of personal information, data stored in a database or files. Security and privacy settings, or some other social services, such as sending friend requests, sending messages or posting on walls. All these resources and services intend to be used by human users on the client computer. User's request for a specific resource or a specific service is sent by the client computer to the web server, which is granted to it if the resource or the service is not protected. In case the resource or the service is CAPTCHA protected, the access is granted to it only after passing CAPTCHA Challenge. The general CAPTCHA architecture is shown in Fig. 1.

The authentication life cycle in CAPTCHA architecture starts from the client by sending a specific request to the web Server. If the request is CAPTCHA protected, the CAPTCHA Image Generation (CIG) generates CAPTCHA image from Image Database, and presents it through CAPTCHA page at the client side to solve it. The state information along with Global Unique Identifier (GUID) of the client and the CAPTCHA solution is stored in the State Information Database (SID) at the server. Storing GUID of the client ensures that only the client that received CAPTCHA page can produce a valid solution through the current cookie. A human user responds to CAPTCHA challenge and the response is passed by the client to the server that call CAPTCHA Verifier Algorithm to verify the authenticity of CAPTCHA solution by comparing the stored

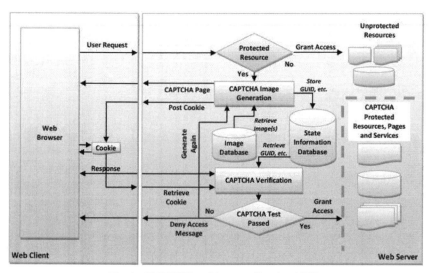

Fig. 1. CAPTCHA architecture (Banday 2009).

Global Unique Identifier (GUID) and the GUID of the client sending the solution. The solution provided by the client is next compared with the solution stored in the State Information Database (SID) and according to the verification result, either the access to the required resource/service is granted or denied. In case of access is denied, an error message is raised up to the client; hence, the CAPTCHA Image Generator (CIG) derives a new CAPTCHA image and the Authentication Life Cycle starts again with a new CAPTCHA cookie. If the client repeatedly fails to solve the number of CAPTCHA challenges, the CAPTCHA system may block the access against this user.

Classical CAPTCHA Generation

The known CAPTCHAs techniques can be classified into three basic families that represent the classical era of CAPTCHA approaches (1) Text-based CAPTCHA which can be designed by carrying out some processing steps on a set of characters, numbers, or words to appear as a distorted and tiling string. (2) Image-based CAPTCHA which can be designed by carrying out some processing steps on a set of objects that appear in a confused image. (3) Audio-based CAPTCHA which can be designed by carrying out some processing steps on a recorded voice that can be heard within jamming and noisy audios. The three classical CAPTCHAs can be detailed as follows:

Text-based CAPTCHA

Text-based CAPTCHA was initially devised by Andrei Broder and his colleagues in 1997 and coined in 2003 by Luis von Ahn et al. (Von Ahn 2003). Text-based CAPTCHAs are presented in the form of an image containing distorted text string as described in Fig. 2 (e.g., characters or numbers) to be identified and typed by the user in a text box provided near the CAPTCHA image on the Web page. To design text-based CAPTCHAs, several design considerations can be applied to the CAPTCHA image in order to present it in a distorted manner to be easy for the user and tough

Fig. 2. Text-based CAPTCHA.

to the Bot software (Bursztein et al. 2011). These considerations can be described as:

1. *Using multi fonts*: Text-based CAPTCHA characters should be written using multiple fonts or font faces.

2. *Using a specified char-set*: The characters and numbers of text-based CAPTCHA should be selected from a specified char-set.

3. *Distortion*: Text-based CAPTCHA should be distorted globally using at-tractor fields.

4. *Blurring*: Text-based CAPTCHA should involve blurring letters in its nature.

5. *Tilting*: Text-based CAPTCHA should involve rotated characters with various angles.

6. *Waving*: Text-based CAPTCHA may be involving rotated characters in a wave fashion.

7. *Complex background*: The design of text-based CAPTCHA should try to hide the text in a complex background to 'confuse' the solver.

8. *Lines*: Adding extra lines to CAPTCHA will prevent the solver from knowing what are the real character segments.

9. *Collapsing*: Remove the space between characters to prevent segmentation by Bot.

The well-known examples of text-based CAPTCHA are:

EZ-Gimpy & Gimpy CAPTCHA were devised at CMU in 2000 (Von Ahn 2004), both CAPTCHAs are shown in Fig. 3. In EZ-Gimpy, the user is presented with an image of a single word, the image has been distorted and a cluttered. textured background has been added to this image (see Fig. 3(a)). The user is asked to recognize this word to pass the test. On the other hand, the improved version is Gimpy that has more difficult variant of text-based CAPTCHA. In Gimpy CAPTCHA, ten words are presented in distortion, clutter (some repeated and overlapped) and in pairs fashion (see Fig. 3(b)), then the user is asked to name and list three of the ten words in the image in order to pass the test.

(a) EZ-Gimpy   (b) Gimpy

Fig. 3. EZ-Gimpy CAPTCHA and Gimpy CAPTCHA.

Baffle CAPTCHAs (Monica and Baird 2003) is a reading-based CAPTCHA that uses random masking to degrade images of non-English pronounceable character strings and asks the user to recognize the masked words as depicted in Fig. 4. Each Baffle text challenge is generated as follows:

1. Generate a pronounceable character string and ensure it is not in the English dictionary.
2. Choose a font from among a large number.
3. Render the character string using the font into an image (without physics-based degradations).
4. Generate a mask image.
5. Choose a masking operation from among add, subtract and difference.
6. Combine the character-string image and mask image using the masking operation.

Fig. 4. Baffle CAPTCHAs.

Scatter CAPTCHAs: In this CAPTCHA, no physics-based image degradations, occlusions or extraneous patterns are performed. Instead, characters of the image are fragmented using horizontal and vertical cuts and the fragments are pseudo randomly scattered by horizontal and vertical displacement (Baird 2005). Scatter CAPTCHA is designed to resist character segmentation attacks. Its challenges are pseudo randomly synthesized images of text strings rendered in machine-print type-faces—within each image, characters are fragmented using horizontal and vertical cuts and the fragments are scattered by vertical and horizontal displacements. This scattering is designed to defeat all methods known to us for automatic segmentation into characters. Figure 5 shows Scatter CAPTCHA.

ReCAPTCHA (Von Ahn 2008) is another text-based challenge based on two different words called 'Control' word and 'unknown' word that are taken from digitalized texts. If the user correctly types the control word, the system assumes he is a human and gains confidence that he will also type the other word correctly. reCAPTCHA test is shown in Fig. 6.

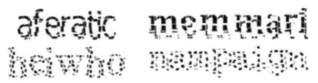

Fig. 5. Scatter CAPTCHA.

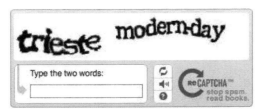

Fig. 6. ReCAPTCHA.

Image-based CAPTCHA

Image-based CAPTCHA is another class of the classical CAPTCHAs which was initially devised by (Blum et al. 2000). In image-based CAPTCHA the user is asked to recognize on specified objects in the presented image to pass the test as depicted in Fig. 7 also the image-based CAPTCHA may be presented to the user as an interactive image such that moving objects from on position to another are seen in the presented CAPTCHA image. Image-based CAPTCHAS depends on some design considerations, such as the use of different patterns or concepts which the user is asked to recognize, like the algorithms for image generation, size, and dimensions of generated CAPTCHA image, the size and types of the images in image database. In addition, image-based CAPTCHA may requires various transformations on the objects to increase the CAPTCHA security, such as scaling, rotation, transparency, color quantization, dithering and adding noise (Raj et al. 2012). The well-known examples of image-based CAPTCHAs can be described as follows:

ESP-PIX CAPTCHA: Detecting visual objects or visual concepts presents a large gap between computer and humans. Image-based CAPTCHAs ask the user to solve a visual pattern recognition challenge or to interpret expressed concepts in the presented image CAPTCHA. ESP-PIX CAPTCHA is the oldest type of image-based CAPTCHA that authenticates humans by recognizing what object is common in a set of images (Blum 2000). The challenge is designed by randomly selecting a specific object or concept and picking four pictures from the database around this object or concept, which are then randomly distorted and presented to the user as a challenge. The test then asks, the user a question like, "What is the word

Fig. 7. Image-based CAPTCHA.

Fig. 8. ESP-PIX CAPTCHA.

that relates to all the images?" and offers some options to choose among them as explored in Fig. 8.

KittenAuth CAPTCHA: Such a type of CAPTCHA is proposed by Oli Warner in (Oli Warner 2015). KittenAuth CAPTCHA is designed by presenting several pictures to the user with many different species of animals and the user must select all the kitten pictures (i.e., required object) to pass the test (Alessandro and Bergadano 2010). For example, in Fig. 9, the user must select all 'tiger' pictures in order to authenticate him as a human, not a Bot. The security of KittenAuth may be enhanced and improved by adding more pictures or asking the user to choose more images.

Bongo CAPTCHA: Another example of image-based CAPTCHA is Bongo. Such a type of CAPTCHA asks the user to solve a visual pattern recognition challenge by presenting two series of block shapes, with series in the LEFT and the other in the RIGHT (Alessandro and Bergadano 2010, Von Ahn 2004). The block shapes in the left series differ from those in the right. After seeing the two left and right series of blocks, the user is presented with a

Fig. 9. KittenAuth CAPTCHA.

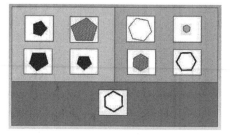

Fig. 10. Bongo CAPTCHA.

single block and asked to determine whether this block belongs to the right series or to the left. The user passes the Bongo challenge if he or she correctly determines the correct series (or side) to which the block belongs. For example, Fig. 10 depicts Bongo CAPTCHA in which the user is asked to know which side does the isolated block belong. Sure, the answer is RIGHT. From the security point of view, Bongo CAPTCHA is vulnerable to the random guessing attacks such that the intruder randomly chooses either the left or the right series. Hence, the random guessing rate is 50 per cent of the time. The matter makes Bongo CAPTCHA insecure although it is friendly to the human user.

Imagination CAPTCHA: Image generation for Internet Authentication is a further image-based CAPTCHA in which the user is asked to solve two tests (Datta et al. 2005). The first test is visualized by presenting a composite image that contains a sequence of eight sub-images, different in size, and then the user is asked to click in the center of a specific sub-image. Sure, this task is simple for the user, but it become a tough challenge to the Bot. In the second test, the selected sub-image is then displayed, enlarged, and controlled composite distortions are applied on it. Then the user is asked to select the word that represents a specific concept expressed by the image from a list of possible options. Figure 11 explores an example of the Imagination CAPTCHA. According to the results in (Datta 2005), the CAPTCHA appears quite effective in its security. However, it requires a human user to solve two distinct challenges. The matter that negatively affects its usability is such that the probability of failure to solve the presented tests increases. In addition, from the given examples, it turns out that the overall screen space required to display the image is very large (800 x 600 pixels).

ARTiFACIAL CAPTCHA: Automated Reverse Turing test using FACIAL features is a further image-based CAPTCHA that exploits human ability to recognize human faces from a presented image (Rui and Liu 2003). ARTiFACIAL is a good example to pass face graphical passwords (Shoba Bindue 2015), such that it works as follows—per each user request, it automatically presents an image with a distorted face embedded in a cluttered and garbled background. The user is asked to first recognize the face and then click on 6 points (4 eye corners and 2 mouth corners) on the face. If the user can correctly identify these points, we can conclude that the user is a human; otherwise, the user is a machine. ARTiFACIAL CAPTCHA is described in Fig. 12. From the usability point of view, 99.7 per cent of users could pass the test in an average time of 14s. However, identifying 6 points of the face contributes to increase in the probability of failure.

Fig. 11. Imagination CAPTCHA.

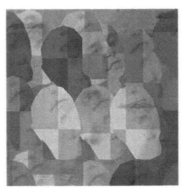

Fig. 12. ARTiFACIAL CAPTCHA.

EasyPIC and MosaHIP are drag and drop CAPTCHAs. The challenge in EasyPIC and MosaHIP are based on the human ability to recognize a specific object displayed by a picture. The two CAPTCHAs schemes are raised up when the user wants to download a specific resource from a specific server online (Alessandro and Bergadano 2010). In EasyPic, the user has to drag the name of the specific resource he/she wants to download and which is expressed in the form of a movable text object rather than a web link, and to drop it on the box according to the picture suggests to do in the page. When the user drops a resource on the indicated right image, that content is sent to the server to download the required resources. Otherwise, a new test is derived with a doubled number of images, which is different from the previous ones. As long as the user fails to pass the test the challenge is increased by duplicating the number of images in which the user is asked to drag the resources and drop in one of them. If the number of failures reaches a specific threshold, then a specific action is taken or the service is blocked. Figure 13 explores an example of EasyPIC CAPTCHA working. The challenge presents two pictures for the user and asks him

Fig. 13. EasyPIC CAPTCHA.

to drag the required resources and drop it on the box showing a specific picture. If the user drops the required resource on the right image's box, then the resource is added to the downloadable resources list on the right and eventually downloaded. On the other side, if the resource is dropped on the other image, a new test with four different pictures is displayed and, if the user keeps failing the test, a new CAPTCHA with eight images is shown to him/her.

Mosaic-based Human Interactive Proof (MosaHIP) is another drag-and-drop image-based CAPTCHA (Alessandro and Sicco 2009, Banday et al. 2009, Ban-day et al. 2015). MosaHIP is an improved version of EasyPIC CAPTCHA from security and usability point of view. Such a type of CAPTCHA exploits the current computer's weakness in image segmentation in the presence of complex background; recognition of specific concepts from noisy background and shape-matching when specific transformations are applied to photos. MosaHIP CAPTCHA challenges the user with a single large image, called mosaic image composed by smaller and partially overlapping photos. These photos are presented according to two different categories—photos representing real and existing concepts and the other photos are fake ones that present artificially a fake and nonexistent concept (Alessandro and Sicco 2009). In MosaHIP challenge, the user is asked to identify a specific photo among real pictures and fake pictures. The real ones are pseudo randomly placed within the mosaic image. These real pictures are presented partially in overlapped and overlaid fashion. In addition, the fake picture are created by filling the background with randomly chosen colors from the color histogram of the real pictures as well as applying some deformation steps (Raj et al. 2012), such as Scaling, Rotation, Transparency, Color Quantization, Dithering, Adding Noise and Rescaling. The user uses the drag-and-drop technique as in EasyPic CAPTCHA through two classes of challenges—the first one is 'concept-based' in which the user is asked to drag the required resources and drop them on the picture's box that expresses a specific concept. The second class is 'topmost' in which the user is asked to drag the required resources and drop them on the picture that presents something existing and laying upon other pictures, not overlapped by anything else. The main issue of MosaHIP CAPTCHA is related to its usability and friendliness for users. Indeed, the usability tests presented in (Alessandro and Sicco 2009) show that 98 per cent of participants could correctly solve the concept-based version, whilst only 80 per cent of participants were able to solve the topmost version. Hence, the concept-based one is the more usable for its nature. Figure 14 presents the concept-based MosaHIP CAPTCHA.

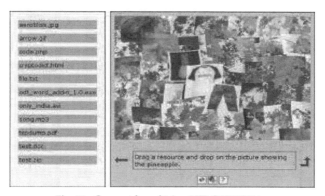

Fig. 14. Concept-based MosaHIP CAPTCHA.

Audio-based CAPTCHA

Another important category of classical CAPTCHAs is audio CAPTCHA (Banday 2009). The first audio-based test was implemented by Nancy Chan and afterwards presented in (Nancy 2002). Such a type of CAPTCHAs depends on computer weaknesses in interpreting the spoken language in the presence of distortion and noisy audio background. The test challenges the user to write digits of a specific string or an identified word which is pronounced at randomly spaced intervals in an audio clip. Both male and female voices are used in common audio CAPTCHAs. The user listens to the clip and is asked to write the content of what he/she heard If the user correctly recognizes all spoken letters or digits, the CAPTCHA is considered passed (Von Ahn 2004) as shown in Fig. 15. For enhancing the diffculty of audio-based CAPTCHA tests, distortion, and additional audio effects must be added to the audio clip, such as reverberation and combination of male and female voices in order to prevent the Automatic Speech Recognition (ASR) systems from solving the audio CAPTCHA. A typical example of audio-based CAPTCHA is reCAPTCHA (i.e., audio version beside text-based CAPTCHA) scheme that is proposed by the

Fig. 15. Audio-based CAPTCHA.

Fig. 16. Two audio CAPTCHA versions.

Carnegie Mellon University. This mechanism also provides users with an audio clip in which eight numbers are spoken by different voices.

ReCAPTCHA (audio version) have been deployed in Google.com and Digg.com (Gao 2010). Another example of audio-based CAPTCHA is introduced in (Sauer et al. 2008), such that the audio CAPTCHA challenge is associated with an image-based CAPTCHA (see Fig. 16). Recent research efforts in audio-based CAPTCHAs is still immature and incomplete since its design has some difficulties, is time consuming and not usable (Bigham and Cavender 2009). Hence improvement studies from the viewpoint of security and usability are still necessary. Recently, some works have started to explore the possible vulnerabilities of existing audio CAPTCHAs and introduced some suggestions to improve them (Tam et al. 2008). Audio CAPTCHA has not received the same level of attention of text and image-based CAPTCHAs because it is intrinsically more difficult, hard to design and requires more hardware to hear the sound. In addition, similar to visual CAPTCHA, an audio CAPTCHA cannot provide full accessibility, since it represents a barrier for audio-impaired users. Therefore, it is currently used as a measure to grant accessibility where visual one cannot (Holman et al. 2007, Schlaikjer 2007).

Next CAPTCHA Generation

The research in CAPTCHAs as a Human Interactive Proof (HIP) still introduces novel CAPTCHA tests to immune online services and resources against a variety of Bot attacks for breaking CAPTCHA system, such as Smuggling Attacks (Manuel 2010), Segmentation and Recognition Attacks (Banday 2009, Gutmann 2006), Relay Attacks (Mohamed et al. 2014) and other classes of attacks which will be discussed in detail in the next section. Some of the recent proposed CAPTCHA approaches are game-based CAPTCHAs, cognitive puzzle-based CAPTCHAs, math-based CAPTCHAs and 3D-based CAPTCHAs. In this section, we will discuss these classes of CAPTCHA approaches in detail.

Game-based CAPTCHA

Existing classical CAPTCHA solutions on the Internet are a major source of user frustration and are breakable. Hence, researchers have started to design and develop a novel CAPTCHA approach based on modern, small, compact games to verify human interaction on web sites. The basic objective is to increase usability and friendliness as well as enhancing security against Bot attacks. We can discuss these approaches below.

PlayThru CAPTCHAs is a new CAPTCHA that is designed to be more simple, more intuitive and more fun than other CAPTCHA options. This type of CAPTCHA is called Dynamic Cognitive Game (DCG) (Mohamed et al. 2014). It challenges the users to solve a simple and precise theme game that is presented through an image (Kuo 2012). The game test in Play Thru CAPTCHA is started by asking user to click and drag the correct objects among a set of objects to drop it in the right position or to make a specific thing or to complete a specific task, such as parking game, animal game, or shape game (Mohamed 2014). For example, the Play Thru CAPTCHA in Fig. 17 asks the user to finish the game by putting the golf ball in the hall in order to be identified as a human, not a Bot. From the usability point of view, PlayThru can be impeded directly on a page or an image as a model. In addition Play Thru has increased conversion rates of 40–60 per cent for its users (Kuo 2012). Also, it is a uniquely elegant solution to the poor user experience presented by traditional text-based CAPTCHAs.

Fun CAPTCHA is another game-based CAPTCHA that is called Tick Tack Toe Game. The test is presented to the user as a set of configurations and he is asked to get number of Xs in a row in each one of the presented

Fig. 17. Play Thru CAPTCHA.

configurations (SwipAds 2014). The user's selection is then submitted to the server and verified to ensure that the submitted solution is from a human rather than a machine. For example, Fig. 18 presents a Fun CAPTCHA which asks the user to get the three Xs in a row in all the three presented configurations to be identified as a human. The main advantages of Fun CAPTCHA are: (1) it is a unique and a fun user interface design; (2) it is very easy to integrate in a specific page; (3) since it is considered a Tick Tack Toe game, it can be solved by anyone; (4) it has a clear and friendly user interface.

GISCHA is a game-based image semantic CAPTCHA which is designed especially for portable and handset devices (Yang et al. 2013). The functionality of GISCHA can be described by using a rolling ball which moves on a two colored square surface, and destination halls with different shapes (Fig. 19a). The user who is able to move the ball to the destination hole shaped as a circle is identified as a human. However, for a computer, it would be tough to understand the meaning of the rolling ball game to make corresponding movement. Moreover even if image processing techniques are developed to break such CAPTCHA to the level where the computer could recognize the meanings of the images and make correct

Get 3 X's in a row in all the 3 games and we'll know your a human.

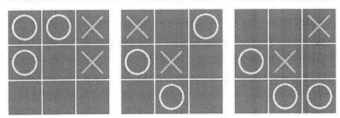

Fig. 18. Fun CAPTCHA.

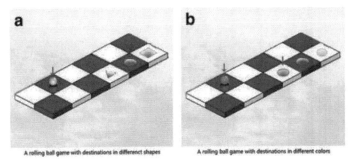

Fig. 19. GISCHA CAPTCHA: (a) a rolling ball game with destination in difierent shapes; (b) a rolling ball game with destination in diffierent colors.

responses, it would be almost impossible for computers to understand the real meaning of the rolling ball game by adding traps and enticing the user to fake destinations (Fig. 19b), in which destinations have the same shape (circular) but are in different colors as in (Yang et al. 2013).

Cognitive Puzzle-based CAPTCHAs

This class of recent CAPTCHAs recognize the human action by presenting the CAPTCHA in the form a cognitive puzzle. The user must think on how to solve the presented puzzle rather than to rewrite specific letters as in text-based CAPTCHAs or to recognize specific images by clicking them as in image-based CAPTCHAs or hear specific characters and then write them down in the textbox. Although puzzle-based CAPTCHAs is a good approach from the security point of view, but some types may bother users and cause frustration. Figure 20 show an example of cognitive-based CAPTCHA that presents four pieces, each to be placed in the appropriate position to constitute and visualize the required shape. If the user is unable to do this task, then the server will consider him as a human rather than a Bot.

The most recent cognitive puzzle-based CAPTCHAs can be discussed as follows:

CAPTCHAStar is a novel puzzle-based CAPTCHA that depends on human cognitive ability to recognize specific shapes within a confused environment (Mauro et al. 2015). The method in CAPTCHaStar prompts the user to use some stars inside a square. The position of these stars changes according to the position of the cursor. The user must move the cursor in the drawable space until the stars aggregate in a recognizable shape. Figure 21 explores stages of solving CAPTCHAStar. From the usability point of view, users are able to complete CAPTCHaStar challenges on an average time of less than 27 seconds with a success rate of more than 90 per cent. In addition, CAPTCHAStar is more resilient against traditional and automated ad-hoc attacks (Gutmann 2006).

Fig. 20. Cognitive puzzle-based CAPTCHA.

(a) A random starting picture.   (b) A sample unsolved challenge.   (c) An almost solved challenge.   (d) A correctly solved challenge.

Fig. 21. The stages of solving CAPTCHAStar challenge.

DnD & DDIM CAPTCHA is another puzzle-based CAPTCHA approach which can be classified as drag-and-drop CAPTCHAs. DnD CAPTCHA is proposed and introduced in (Desai 2009). In DnD CAPTCHA, the user is asked to drag each character of a specific word that appears in distorted nature and drop it in the right position of this character. In other words, the user cannot type characters on the text box for CAPTCHA test as in text-based CAPTCHAs classes, but in reverse, the user has to just simply drag and drop character blocks into their appropriate blank blocks (Samruddhi and Mishra 2015). Figure 22 presents the DnD CAPTCHA interface. In this DnD CAPTCHA BOT or malicious computer program may be able to know the distorted word in the presented test through Optical Character Recognition (OCR) techniques, but it cannot pass the test because dragging and dropping characters into the appropriate blank boxes is a very tough task for the Bot. Another similar drag-and-drop CAPTCHA approach is drag-n-drop interactive masking CAPTCHA or DDIM-CAPTCHA (Samruddhi and Mishra 2015, Ye et al. 2013), that uses masking techniques and some interaction principles to design a challenge in the form CAPTCHA to resist third-party human attacks. In general, the DDIM CAPTCHA maintains the basic prerequisites of CAPTCHA with added interaction rules and masking properties of the traditional text-based scheme.

TapCHA is a novel gesture-based CAPTCHA which is designed for use in ubiquitous computing environments, such as smartphones and tablet devices. Such a type of puzzle-based CAPTCHA is proposed by Nan Jiang in his paper (Nan and Tian 2013). The main idea on which TapCHAs design system is constructed is based on a hybrid challenge that combines

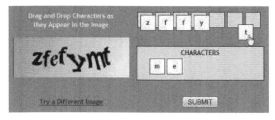

Fig. 22. Drag-and-drop CAPTCHA.

two puzzles—text recognition puzzle and shape motion puzzle. The two puzzles are based on the advantages of human cognitive capabilities. The challenge, which contains two processing phases, involves text and object recognition and motion puzzles to enhance the security of interactive CAPTCHAs while keeping the response process simple, fast and intuitive on smart devices. The first phase contains a text-recognition-based reading comprehension task where a user needs to recognize the text presented on the screen and understand its meaning as it provides instruction for the next challenge. The second phase combines some shape-and-motion puzzles based on the context presented in the first phase. The general architecture of TapCHA is shown in Fig. 23. The motion puzzle in TapCHA is the core challenge phase since it identifies how users interact and solve the CAPTCHA on smart devices.

In phase one, the instructions of what the user is asked to do are presented in a text-box in a similar way as in letters in text-based CAPTCHAs shown in Fig. 24(a). The puzzle here is that the user needs to perceive the meaning of distinguished words which explain what he is asked to do within the context of the instructions. Certainly, Bots may be able to identify distorted characters of the distinguished words within the context by using advanced AI and image-processing techniques. However, they cannot easily understand the meaning and semantic of a statement formed by these words. Hence, this technique will improve security compared to the pure clear text instruction (Nan and Tian 2013). In phase two, the motion challenge is provided based on the instructions

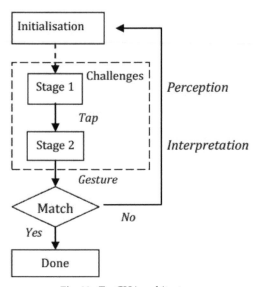

Fig. 23. TapCHA architecture.

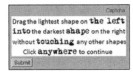 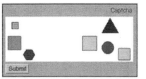

(a) TapCHA Instruction   (b) TapCHA Motion
(Phase 1)   (Phase 2)

Fig. 24. TapCHA phases.

above where various transformed shapes are drawn on the screen and the user is asked to move one of them according to the instructions. Figure 24(b) explores this phase. The TapCHA approach is similar to the shape CAPTCHA design that is proposed by Rusu and Docimo in (Rusu and Docimo 2009), but TapCHA has an advantage since motion response can be easily achieved by using gestural interactions on touch screens and mouse-based interactions on conventional screens. In addition, TapCHA is more protected against OCR attacks (Xu et al. 2012).

M CAPTCHA is a new human interactive proof approach (HIP) to prevent Bot spam by asking user to draw a randomly generated pattern as a graph puzzle. M CAPTCHA (ENVATO MARKET 2013) is targeted for touch-screen device, such as smartphones and tablet devices that are based on android operating system (Reddy 2014). M CAPTCHA works well on every major desktop and mobile browser. For a device without the touch screen, the pattern can be drawn using mouse. M CAPTCHA puzzle is based on a graph G which consists of nine vertices connected via different lines and the arrows indicate the order of connected vertices. For solving M CAPTCHA, the user will be asked to draw a similar pattern as shown in the CAPTCHA image, by connecting the vertices in the same order as shown in the CAPTCHA image either by touching the screen in smart devices or using the mouse in laptops. Figure 25 explores M CAPTCHA example in which the user is asked to draw ' Z' pattern example according

M CAPTCHA

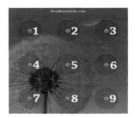 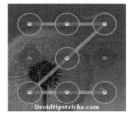

Fig. 25. M CAPTCHA (Z pattern example).

to the vertices sequence 1-2-3-5-7-8-9. M_CAPTCHA is featured as an outstanding CAPTCHA since it targets the touch-screen device so that drawing on the touch screen is far easier than typing (more usable). M CAPTCHA is fully customizable, such that the user can change the color of arrows, of vertices, or background, etc. in addition, M_CAPTCHA is easy to integrate and is well documented.

Math-based CAPTCHA

An attractive class of recent CAPTCHAs is math-based CAPTCHA (Castro et al. 2010). Such a CAPTCHA can be used to distinguish human actions by using math operations, such as addition, subtraction, multiplication, or other mathematical problems. It can generate an image that displays a simple mathematical operation between two random integers. The user is asked to enter the result of the math operation to pass the test as described in Fig. 26. Math CAPTCHA is used commonly during digital online forms. Some online services may ask the user to fill some personal fields, such as his/her name, his/her email/and other required information. Then user is asked to solve the math CAPTCHA to prove that the form is submitted by a human, not a Bot. Some online services include Quantum Random Bit Generator Service (QRBGS) (Stevanovic et al. 2008). The service has begun as a result of an attempt to fulfil the scientists' needs for quality random numbers that are used in many scientific simulation problems but has now grown to a global (public) high-quality random numbers service. Once a specific user login this web service, he/she can use one of the many software client libraries to connect to the server and download the random numbers provided by them when needed. In order to protect this service from vandalism or denial of service, a complex mathematical-based CAPTCHA called QRBGS CAPTCHA is developed (Castro et al. 2010). The uniqueness of such math CAPTCHA is that it supposes that all users have advanced knowledge in mathematics since it provides complex tests based on solving Linear and differential equations apply partial differentiation on specific mathematical terms. Figure 27 explores an example of the QRBGS CAPTCHA.

Another usable class of math-based CAPTCHA is dice CAPTCHA. Dice CAPTCHA is a good example of mathematical randomness (Wilkins 2009) which uses dice faces numbers to visualize a mathematical challenge CAPTCHA to prevent spam Bots to access or submit digital online forms. Users often find some difficulty in identifying the obscured and distorted words that a classical text-based CAPTCHA uses for digital online form of submissions. Dice CAPTCHA is a novel way to reduce this difficulty while still helping to prevent Bot spam malicious activities (Jones 2015). Dice CAPTCHA comes in two versions to recognize the human action, Homo-sapiens Dice and All-the-rest Dice. In the first version (e.g., Homo-

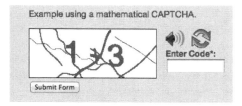

Fig. 26. Math-based CAPTCHA.

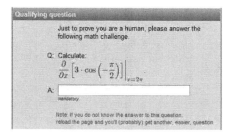

Fig. 27. QRBGS CAPTCHA.

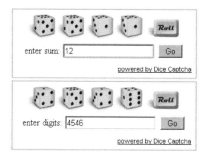

Fig. 28. Two versions of dice CAPTCHA.

sapiens Dice), the user is asked to roll the dice and enter the sum of all dots that appear in all the dices, on the other hand. In All-the-rest-Dice version, the user is asked to enter sequentially the count of dots that appears in each dice respectively. The two versions of dice CAPTCHA are explored in Fig. 28.

3D-based CAPTCHA

3D CAPTCHA is another rising CAPTCHA approach that is based on human imagination and spatial perspective with the ability to recognize real objects in the context of random graphical information (Rolko and Slovakia 2010). The main idea in 3D CAPTCHA is rotation of a special 3D model and finding the correct position of rotation by which the right observation is detected. 3D model-based CAPTCHA can be created from any 2D picture through the following steps, First, the 2D picture is divided into several parts as triangles, which are then randomly projected into 3D space according to a certain rule. Parts projected spatially this way generate together a 3D model, so that the model looks like the original 2D image from a single observation point. The meaning of 3D model observed from any other observation point is incomprehensible to a human. The task of a user is to rotate the model to find the right observation point (i.e., right rotation angle of the 3D model-based CAPTCHA) and solve the CAPTCHA successfully. This concept is based on the fact that the meaning of the model is recognizable from the right observation point only by a human since this task is very hard for the machine. 3D text-based CAPTCHA (Montree and Phimoltares 2010) is a good example of 3D CAPTCHA. The CAPTCHA text consists of alphanumeric characters which are letters and numbers. The text is composed of six characters, each of which has its own axis and rotation angle. Each character is rotated in a certain angle ranging from –45 degree to 45 degrees by using a standard randomization function. The general 3D CAPTCHA challenge is presented to the user through a specific interface after applying ten sequence functions to the alphanumeric letters of CAPTCHA which are (Montree and Phimoltares 2010) rotation, overlapping, adding lines, adding distributed noise, coloring characters, coloring background, scaling, using variety of fonts, using distinguished characters, and using background texture. Figure 29 explores the 3D text-based CAPTCHA with several distorted and noisy backgrounds. The literature introduced three types of 3D CAPTCHAs, which are discussed below.

3D Animation CAPTCHA is another version of 3D-based CAPTCHA tests which is based on the recognition of moving 3D objects in videos frames (Jing-Song 2009). Such a type challenges the user with continued animated frames and CAPTCHA is still shown in the form of characters

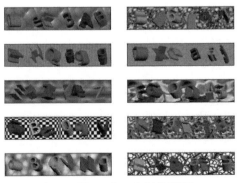

Fig. 29. 3D text-based CAPTCHA.

for users to recognize, but characters are hidden in 3D animation. The animation technique used in 3D Animation CAPTCHA is a very hard obstacle against Bot software to identify the content of CAPTCHA in each frame. Computer programs using the OCR method cannot solve this difficulty.

AniCAP (Yang-Wai and Susilo 2011) is another 3D-animated CAPTCHA technique which is based on motion parallax. This capitalizes on the inherent human ability to perceive depth from the apparent difference in motion of objects located at different distances from a moving viewpoint. Foreground characters and background characters in AniCAP are placed at different depths in the 3D scene. Thus, from the point of view of a moving camera, humans can distinguish the main characters in the foreground against the background characters, because the foreground characters will appear to move at different rates as compared to background characters. AniCap is more resistant against segmentation and recognition attacks. A similar secure motion parallax-based 3D CAPTCHA is Stereoscopic 3D CAPTCHA or STE3D-CAP (Willy et al. 2010) (Yang-Wai and Susilo 2012). The main idea in STE3D-CAP is to produce a stereo pair. Two images of the distorted 3D text objects generated from two different camera/ eye viewpoints are presented to a human user's left and right eyes, respectively. When the two images are supplied to hardware capable of displaying stereoscopic 3D images, the resulting CAPTCHA can easily be solved by a human, but it will be tough on computers.

Sketcha CAPTCHA is another example of 3D CAPTCHA that is based on line drawing of a variety of 3D Objects in different orientation (Steven 2010). The 3D objects are selected from a large database and all the images are presented from random viewpoint angles, The CAPTCHA is presented to the user as a set of 3D sketched images, and the user must rotate all the images till to make each of them into an upright orientation. This task generally requires understanding of the semantic content of each sketched image, which is believed to be more difficult for Bot software to perform. Figure 30 explores such a type of 3D CAPTCHA. The user is asked to pass the test by choosing the upright orientation angle for each image. A similar CAPTCHA technique is Googles CAPTCHA that is based on image orientation, such that the users are presented a set of images that have to be rotated to align them in an upright position (Gossweiler et al. 2009). Another 3D CAPTCHA approach called Avatar CAPTCHA (Souza 2012) that asks the user to identify 3D gray faces images among a set of 12 grayscale images comprised of a mix of human and avatar faces. The main advantages of 3D CAPTCHA are it is based on human imagination; it is fully automated; it depends on unlimited source of images such that the image can be used and projected into 3D space; and with increasing

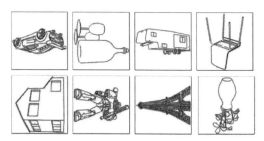

Fig. 30. Sketcha CAPTCHA (user is asked to rotate eight images until upright orientation state).

experience of 3D CAPTCHA test, it can be solved faster than the traditional text-based CAPTCHA system. On the opposite side, the flash-based design 3D CAPTCHAs are not appropriate for all devices and require relatively high demands on connectivity with the user (examples, are about 500 KB in size) (Rolko et al. 2010).

Attack Models and Breaking Techniques

Whether a CAPTCHA system is based on text, image, audio, game, puzzle or 3D CAPTCHA-based design, some of the categorized attacks can be launched by intruders and malicious users to penetrate whatever CAPTCHA system is used. In this section, we list some of the classical and known attacks that target CAPTCHA system and discuss the key tools which the intruders may use to break the used CAPTCHA technique. Some of these attacks can be categorized as follows:

Brute Force Attacks

Brute force attack is a common attack model that targets some CAPTCHAs, such as Asirra (Elson 2007). When the CAPTCHA test is based on a limited number of solutions (characters or images), then, the intruders can use this vulnerability to automatically attack this CAPTCHA by exhaustively trying answers at random or according to a selected sequence. For example, a 4-digit-based CAPTCHA would have 10000 possible solutions to guess (Ball 2012).

Signal Processing Attacks

Signal processing attacks are common attacks in image and audio-based CAPTCHAs (Ball 2012). Although the disorder, noise and perturbations that are commonly used to obscure and confuse image and sound CAPTCHAs are possible tasks for the machine to do and a tough task to reverse. But the intruders can solve image CAPTCHAs by removing the noise and distortion via Optical Character Recognition (OCR) technique (Bursztein et al. 2011) or via mathematical heuristics and machine learning algorithms (Mauro et al. 2015), such as content-based image retrieval (CBIR) methods (Banday et al. 2009). In addition, he/she can break audio CAPTCHA, using some machine learning algorithms, such as Ada Boost, Support Vector Machine and K-Nearest Neighbor (K-NN) (Tam et al. 2008) in order to recognize the pronounced characters within a noisy voice environment.

Relay Attacks

Relay attacks are a special kind of man-in-the middle attacks in which malicious users outsource or relay the CAPTCHA test to other unsuspecting or paid users rather than answering the test themselves (Olumide 2015). This attack process is known as CAPTCHA relay attack. Certainly, the real goal of CAPTCHAs is immunity against relay attack and is not to prevent Bots, but to keep out the malicious users who use Bots and scam programs to automate the fraudsters' interests. Scammers have therefore invented a creative way to provide their Bots with CAPTCHA-solving capabilities by outsourcing the CAPTCHA-solving task to humans. For example, when a Bot is faced with a CAPTCHA system as access control, it might place the CAPTCHA on the access page for a porn site, and the next visitor to that site solves the CAPTCHA for the Bot. Next, the Bot uses the CAPTCHA solution that is provided by humans who access on porn sites to access the targeted online resource or service. Depending on an under-ground CAPTCHA-breaking service of Amazons, Mechanical Turk (Bursztein et al. 2010) is a good example of such a type of CAPTCHA attacks, such that the workers in Mechanical Turk are asked to solve more than 318,000 CAPTCHAs issued from the 21 most popular CAPTCHA schemes (13 image schemes and 8 audio schemes). The result showed that image CAPTCHAs have an average solving time of 9.8 seconds and three-person agreement of 71.0 per cent, and audio CAPTCHA is much harder with an average solving time of 28.4 seconds, and three-person agreement of 31.2 per cent.

Smuggling Attack

Smuggling attack (Manuel 2010) is a novel CAPTCHA attack in which the intruders inject spoofed CAPTCHA challenges to automatically replay an identified online task, such as mail registration, login social application, such as facebook applications, sending message or commenting on message/photo or sending friend requests, etc. Smuggling attack behavior is controlled by an attacker who injects and sets up malware software on the victims host. The scenario of smuggling CAPTCHA attack starts by the user first performing an online task that the attacker wishes to postpone (for example, sending a request to a specific web server, clicking a button on a web page, or login a specific service). The malware on the victim's host intercepts the request and locally stores all information necessary to replay the request later on. This malware component then retrieves a CAPTCHA challenge from the attacker's server as a spoof on. The attacker then forwards that challenge that needs to be solved in order to perform the desired action. He could, for example, forward a CAPTCHA that was encountered during the registration of an e-mail account. Once the user has solved the challenge, the malware replays the intercepted request from the stored information in automated way. Smuggling CAPTCHA attack is considered lightweight attack since the victim is conned into believing that the CAPTCHA is displayed by the legitimate online service. Hence the user has no easy way to distinguish between an authentic and a smuggled-in CAPTCHA.

DeCAPTCHA Pipeline Attack

DeCAPTCHA pipeline attack is a CAPTCHA breaker technique that can be used to break text-based CAPTCHA (Bursztein et al. 2011). Five stages are performed on a specific text CAPTCHA to solve it:

- *Preprocessing*: In this rst stage, the Captcha's background is removed using several algorithms and the Captcha is binarized (represented in black and white) and stored in a matrix of binary values. Transforming the Captcha into a binary matrix makes the rest of the Decaptcha pipeline easier to implement.

- *Segmentation*: In this stage, Decaptcha seeks to segment the Captchas using various segmentation techniques, such as CFS (Yan and El Ahmad 2008) (Color Filling Segmentation) which depends on a paint bucket flood filling algorithm in its work. CFS is considered the default segmentation technique as it allows to segment the Captcha letters even if they are tilted and over-lapped.

- *Post-segmentation*: At this stage, the segments resulted in the last stage are processed individually to make the recognition easier. During this phase, the segment sizes are always normalized.

- *Recognition*: In this stage, a training mode is used to learn the classifier what each letter looks like after the CAPTCHA has been segmented. In testing mode, the classifier is used in predictive mode to recognize each character individually.

- *Post-processing*: During this stage the classifiers output is improved and enhanced when possible by using spell-checking technique to increase the accuracy and precision of the output.

The effciency of DeCAPTCHA is evaluated against 15 real text-based Captchas from Authorize, Baidu, Blizzard, Captcha.net, CNN, Digg, eBay, Google, Megaupload, NIH, Recaptcha, Reddit, Skyrock, Slashdot, and Wikipedia. The results explored that DeCAPTCHA could break 7 CAPTCHA from 15 with high success rate, but Google and Re-CAPTCHA were more resistant to DeCAPTCHA breaker.

Vidoop CAPTCHA Attack

Vidoop CAPTCHA attack is special kind of image-based CAPTCHA attack that could break Vidoop CAPTCHA (Michele and Jacob 2009). It is described previously in Fig. 6. Vidoop CAPTCHA uses images of objects, animals, people or landscapes instead of distorted text to distinguish a human from a computer program. Vidoop challenge image consists of a combination of pictures representing different categories. Each picture is associated with a letter which is embedded in it. In order to pass the challenge, the user is asked to report the letters corresponding to a list of required categories. The breaking algorithm performs three phases to break Vidoop image-based CAPTCHA. The first phase is test image processing in which the single pictures are isolated and the corresponding character region of each picture is extracted. The second phase is image category recognition in which the images are classified according to the categories required by the test using nearest neighbor classifier (Tam et al. 2008). Finally, the third phase is character recognition in which the CAPTCHA breaker is able to extract the characters corresponding to the images classified as being part of the required categories. The experimental result of this attack proved that Vidoop CAPTCHA is not secure, such that the breaking algorithm is able to beat and break 9 challenges in one hour (Michele and Jacob 2009).

Teabag 3D Attack

Teabag 3D attack CAPTCHA is a special type of 3D CAPTCHA such that its design is based on three-dimension space as described in Fig. 31. Such a type of 3D CAPTCHA is characterized with some properties:

- The 3D CAPTCHA test appears on a grid in 3D space as depicted in Fig. 31
- The challenge is presented as Four characters
- Only upper case letters and digits are used in the challenge
- The characters are very close to each other
- Each challenge appears to be generated from slightly different viewpoints

– There are small variations in the grid direction and the shape of background cells between the presented challenges.

The attack model for breaking Teabag 3D CAPTCHA (Nguyen et al. 2011) can be described into four phases as depicted in Fig. 32:

– *Pre-processing*: In this phase the breaker algorithm identifies the regions that involve characters and digits of the challenge and extract the key features that help to separate each character individually. This phase is achieved by performing four activities: (1) adaptive binarization. (2) side surface identification. (3) front surface identification. (4) extracting characters. The functionality of each activity is discussed in detail in (Nguyen et al. 2011).

– *Segmentation*: In this phase, the breaker algorithm decomposes the image into sub-images which only contain single characters. Number of segmentation techniques are used to obtain a set of possible splitters for the characters. The best splitters were then selected from this candidate set.

– *Post-processing*: In this phase, some of post-processing steps are performed to ensure optimal character recognition accuracy, such as: (1) combining the pixels of side and front surfaces to improve image quality of characters. (2) de-skewing and straightness of characters using de-skewing angle that was calculated based on the four extreme corners of CAPTCHAs grid. (3) resizing characters to help the optical character recognition OCR to recognize each character sequentially. (4) refining characters by removing noise and smoothen the character's borders to increase the accuracy of character recognition.

– *Characters recognition*: In this phase, the optical character recognition (OCR) technique is used to identify the presented characters in Teabag 3D CAPTCHA challenge.

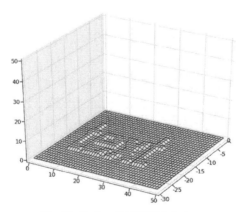

Fig. 31. Teabag CAPTCHA design.

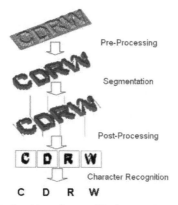

Pre-Processing

Segmentation

Post-Processing

Character Recognition

Fig. 32. Breaking phases of Teabag 3D CAPTCHA.

Security Principles for Designing Secure CAPTCHAs

Analyzing the strength and security of authentication system, such as pass-words (Chanda 2016) concluded that passwords should be long, involve combination of letters, digits, symbols, and should be unguessable by the same way as CAPTCHA is designed according to a set of robustness and security considerations. This section presents some of security design principles to build a strong CAPTCHA system against CAPTCHA attacks or breaking algorithms. To design a strong and secure text and image-based CAPTCHA system, the presented challenge must be very tough against segmentation algorithms and recognition algorithms. The following principles lead to more resistant CAPTCHA system (Bursztein et al. 2011):

1. *Randomize the CAPTCHA length*: Dont use a fixed length while designing CAPTCHA challenge because it gives too much information to the attacker.

2. *Randomize character/image size*: The randomness is very good approach to design hard CAPTCHA. In text CAPTCHA, you should use several font types and several font sizes for reducing classifier learnability and accuracy. For image CAPTCHA, several colors, several confused and distorted pictures, and several picture sizes in random way are used for misleading the classifier algorithm.

3. *Wave the CAPTCHA*: Waving the CAPTCHA enhances the hardness of finding segment points in case of collapsing and helps mitigate the risk of the attacker to remove the added lines.

4. *Use complex background*: Although use of complex background that contains dots, shapes and lines confuses the real text and prevents the breaker from segment characters, previous works Bursztein et

al. (2011) have explored that usually this type of defence has limited security and represents security vulnerability. Instead, the good security principle is to use anti-pattern which is a proper way to deal with CAPTCHA schemes that use random backgrounds but a finite number of colors. In addition, a related approach to the complex background techniques is to use colors that are perceived as very different by humans but are in reality very close in the RGB spectrum. Hence, this technique confuses the breaker to distinguish characters of presented CAPTCHA. regarding image-based CAPTCHA security against object matching and segmentation can be improved by distorting CAPTCHA images by application of transformations, like scaling, rotation and transparency. This makes restoration of images to original form a very tough task for breakers (Banday et al. 2007).

5. *Do not use complex charset*: Although using a larger charset somewhat impacts the classifier's learnability and accuracy, but, the charset of CAPTCHA that does not involve confusing characters (e.g., i/j or 1/i/l or 5/s, etc.) make it easy for humans to solve it.

6. *Adding noise*: An efficient mechanism to confuse the segmentation attacks is to add random noise to the image or text CAPTCHA to make it more resistant against de-noising algorithms, such as erode (Wilkins 2010).

7. *Use large cross lines*: Using lines that are not as wide as the character segments gives an attacker a robust discriminator and makes the line anti-segmentation technique vulnerable to many attack techniques, including de-noising, projection-based segmentation. It is an important principle to make the lines pass over CAPTCHA characters to confuse the breaker algorithm to perform segmentation attack. Also, keep the angle of the line on par with the character segments, otherwise the line slope will be used as a discriminator by the attacker.

8. *Use collapsing*: Having characters collapsed either by removing the space between characters or by tilting them is a recommended principle to prevent the segmentation attack. It is also advised to not use too aggressive collapsing to keep the CAPTCHA usable for humans.

9. *Use curser, clicking, and dragging and dropping techniques*: Although solving CAPTCHA, especially cognitive puzzle-based CAPTCHA using curser motion (Mauro et al. 2010), clicking and move objects (Nan and Tian 2013), or dragging and drop (Desai et al. 2009, Ye et al. 2013) is an easy and usable task for a human, but designing CAPTCHAs using these techniques makes breaking CAPTCHA a very hard problem for the Bot attacks, such as ad-hoc attacks (Gutmann et al. 2006) and relay attacks (Mohamed et al. 2014).

10. *Use Simple Game*: Designing CAPTCHA challenge based on playing funny simple game (Yang et al. 2013) is a very good technique to make the CAPTCHA challenge more resistant against automated attacks. Exploiting the limited capabilities of computer vision to develop CAPTCHA system based on solving game puzzle is a good option to resist against malicious Bots.

Conclusion

Completely Automated Public Turing Test to Tell Computers and Humans Apart (CAPTCHA) is an important and common mechanism that represents the Human Interactive Proof (HIP) approach. CAPTCHA is a challenge-response test to differentiate between behaviors and actions that are performed via human and the automated actions that are undertaken by automated software (Bots). Immunizing and protecting web services and resources require developing of a secure CAPTCHA system to work as an access control and authentication system against malicious Bots. Several resources and services ask the party who aims to access the services to prove that he is a human rather than a Bot. This paper introduces a review study around two generations of CAPTCHA approaches (i.e., classical and modern generations). In addition, it discusses a set of malicious attacks and CAPTCHA breaker techniques that threaten any CAPTCHA system. Finally, this review is ended with some key security principles that should be followed to develop and design a more secure and robust CAPTCHA system against Bot attacks.

References

Ahn, V., L. Blum, M. Hopper, J. Nicholas and L. John. (2003). CAPTCHA: Using hard AI problems for security. pp. 294–311. *In*: The Proceedings of the International Conference on the Theory and Applications of Cryptographic Techniques. Springer.

Alessandro, B. and S. Sicco. (2009). Preventing massive automated access to web resources. Computers & Security 28(3): 174–188.

Alessandro, B. and F. Bergadano. (2010). Anti-bot strategies based on human interactive proofs. pp. 273–291. *In*: Handbook of Information and Communication Security. Springer Berlin Heidelberg.

Baird, H. S., M. A. Moll and S. Y. Wang. (2005). Scatter type: A legible but hard-to-segment CAPTCHA. pp. 935–939. *In*: The Proceedings of 8th International Conference on Document Analysis and Recognition (IEEE Computer Society, Washington 2005), doi:10.1109/ICDAR.2005.205.

Ball, G. B. (2012). Strengthening CAPTCHA-based web security. First Monday Journal 17(2): Feb 2012.

Banday, M. T. and N. A. Shah. (2011). A study of CAPTCHAs for securing web services. International of Secure Digital Information Age (ISDIA) 1(2).

Bigham, J. P. and A. C. Cavender. (2009). Evaluating existing audio CAPTCHAs and an interface optimized for non-visual use. pp. 1829–1838. *In*: Proceedings of ACM CHI 2009 Conference on Human Factors in Computing Systems.

Blum, M., von Ahn and J. Langford. (2000). The CAPTCHA Project, Completely Automatic Public Turing Test to Tell Computers and Humans Apart, School of Computer Science, Carnegie-Mellon University, http://www. captcha. net.

Bursztein, E., S. Bethard, C. Fabry, J. C. Mitchell and D. Jurafsky. (2010). How good are humans at solving CAPTCHAs? a large scale evaluation. In Security and Privacy (SP), IEEE Symposium on IEEE pp. 399–413.

Bursztein, E., M. Martin and J. Mitchell. (2011). Text-based CAPTCHA strengths and weaknesses. pp. 125–138. *In*: Proceedings of the 18th ACM Conference on Computer and Communications Security. ACM.

Castro, H., C. Javier and A. Ribagorda. (2010). Pitfalls in CAPTCHA design and implementation: The Math CAPTCHA, a case study. Elsevier, Computers & Security 29(1): 141–157.

Chanda, K. (2016). Password Security: An analysis of password strengths and vulnerabilities. I.J. Computer Networks and Information Security (IJCNIS), MECS, 8(7): 23–30.

Datta, R., J. Li and J. Z. Wang. (2005). Imagination: a robust image-based CAPTCHA generation system. pp. 331334. *In*: The Proceedings of 13th ACM International Conference on Multimedia (MULTIMEDIA05) (ACM Press, New York 2005).

Desai, A. and P. Patadia. (2009). Drag and drop: A better approach to CAPTCHA. India Conference (INDICON), 2009 Annual IEEE, Gujarat, pp. 1–4.

Elson, J., J. R. Douceur, J. Howell and J. Saul. (2007). Asirra: a CAPTCHA that Exploits Interest-Aligned Manual Image categorization. pp. 366–374. *In*: The Proceedings of ACM Conference on Computer and Communications Security.

ENVATO MARKET. (2013). 'MCaptcha' (Online), http://codecanyon.net/item/m-captcha/4399437.

Gao, Haichang, Liu, Honggang, Yao, Dan, Liu, Xiyang and Aicke-lin, Uwe. (2010). An audio CAPTCHA to distinguish humans from computers. Electronic Commerce and Security (ISECS), 2010 Third International Symposium on. IEEE.

Gossweiler, R., M. Kamvar and S. Baluja. (2009). Whats up CAPTCHA?: a CAPTCHA based on Image Orientation. *In*: The Proceedings of 18th International Conference on World Wide Web, Madrid, Spain.

Gutmann, P., D. Naccache and C. Palme. (2006). CAPTCHAs: humans vs. Bots. IEEE Security & Privacy 6(1): 68–70.

Holman, J., J. Lazar, J. H. Feng and J. D. Arcy. (2007). Developing usable CAPTCHAs for blind users. Proceedings of the 9th International ACM SIGACCESS Conference on Computers and Accessibility pp. 245246.

Jing-Song, C. (2009). A Captcha implementation based on 3D Animation. Multimedia Information Networking and Security,. MINES'09. International Conference on. Vol. 2. IEEE.

Jones, S. (2015). Dice CAPTCHA (Online), http://codecanyon.steve-j.co.uk/dice-captcha/.

Kanika, S. and R. S. Chadha. (2013). Captcha generation for secure web services. International Journal of Engineering and Innovative Technology (IJEIT) 2(10): 168–171.

Kuo, I. (2012). Beat CAPTCHA by Playing Games with PlayThru (online) http://www.gamification.co/2012/11/19/beat-captcha-by-playing-games-with-playthru/.

Manuel, E. (2010). Captcha smuggling: hijacking web browsing sessions to create Captcha farms. *In*: The Proceedings of the 2010 ACM Symposium on Applied Computing. ACM.

Mauro, C., C. Guarisco and R. Spolaor. (2016). CAPTCHaStar A novel CAPTCHA based on interactive shape discovery. International Conference on Applied Cryptography and Network Security, Springer, pp. 611–628.

Michele, M. and J. Jacob. (2009). Breaking an image based CAPTCHA. Technical Paper submitted to the Department of Computer Science, Columbia University, USA, Springer term, Available at www. cs. columbia. edu/ mmerler/project/FinalReport. Pdf.

Mohamed, M., N. Sachdeva, N. Georgescu, S. Gao, N. Saxena, C. Zhang and W. B. Chen. (2014). A three-way investigation of a Game-CAPTCHA: Automated attacks, relay attacks, and usability. pp. 195–206. *In*: Proceedings of the 9th ACM Symposium on Information, Computer and Communications Security, ACM.

Monica, Ch. and H. S. Baird. (2003). Baffle text: A human interactive proof. DRR 5010: 305–316.

Montree, I. and S. Phimoltares. 2010. 3D CAPTCHA: A Next Generation of the CAPTCHA. Information Science and Applications (ICISA), 2010 International Conference on. IEEE, pp. 1–8.

Nan, J. and F. Tian. (2013). A novel gesture-based CAPTCHA design for smart devices. *In*: The Proceedings of the 27th International BCS Human Computer Interaction Conference. British Computer Society.

Nancy, C. (2002). Sound oriented CAPTCHA. *In*: The Proceedings of the 1st Work-shopon Human Interactive Proofs, Xerox PaloAlto Research Center.

Nguyen, V. D., T. W. Chow and W. Susilo. (2012). Breaking a 3D-based CAPTCHA scheme. pp. 391–405. *In*: The Proceedings of the International Conference in Information Security and Cryptology-ICISC 2011, Springer.

Oli Warner, 'KittenAuthh' (online) https://thepcspy.com/kittenauth/#3, (Accessed Sep./2015).

Olumide, B. L. (2015). Mitigating CAPTCHA relay attacks using multiple challenge-response mechanism. Online: http://www.thefreelibrary.com/Mitigating CAPTCHA relay attacks using multiple challenge-response-mechanism.

Proffitt, B. (2013). Next-Generation Search: Software Bots Will Anticipate Your Needs (online), http://readwrite.com/2013/05/02/future-of-search-software-bots-anticipate-your-needs.

Raj, S. Benson Edwin, V. S. Jayanthi and V. Muthulakshmi. (2012). A novel architecture for the generation of picture based CAPTCHA. International Conference on Advanced Computing, Networking and Security. Springer Berlin Heidelberg, pp. 568–574.

Reddy, A. (2014). Android Pattern Lock Ideas With Hardest Pattern Lock Tips. Online: http://www.droidtipstricks.com/pattern-lock-ideas-hard-cool-common/.

Rolko, J. and Z. Slovakia. (2010). 3D CAPTCHA Captcha-based on Spatial Perspective and Human Imagination", http://3dcaptcha.net/documents/3D captcha.pdf.

Rui, Y. and Z. Liu. (2003). Artifacial: Automated reverse turing test using facial features. pp. 295298. *In*: The Proceedings of 11th ACM International Conference on Multimedia, ACM, New York, MULTIMEDIA03.

Rusu, A. and R. Docimo. (2009). Securing the Web using human perception and visual object interpretation. pp. 613–618. *In*: The Proceedings of 13th International Conference on Information Visualisation, IEEE.

Samruddhi, B. and S. Mishra. (2015). A survay on CAPTCHA technique based on drag and drop mouse action. International Journal of Technical Research and Applications (IJTRA) 3(2): 188–189.

Sauer, G., H. Hochheiser, J. Feng and J. Lazar. (2008). Towards a Universally usable CAPTCHA. *In*: Proceedings of the Symposium on Accessible Privacy and Security, ACM Symposium On Usable Privacy and Security (SOUPS'08), Pittsburgh, PA, USA.

Schlaikjer, A. (2007). A dual-use speech CAPTCHA: Aiding visually impaired web users while providing transcriptions of audio streams. LTI-CMU Technical Report-, Carnegie Mellon University, pp. 07–014.

Shoba Bindue. (2015). Secure usable authentication using strong pass text words. I.J. Computer Networks and Information Security, MECS, 7(3): 57–64.

Souza, Darryl D., Phani C. Polina and Roman V. Yampolskiy. (2012). Avatar Captcha: Telling computers and humans apart via face classification. Electro/Information Technology (EIT), IEEE International Conference on. IEEE pp. 1–6.

Stevanovic, R., G. Topic, K. Skala, M. Stipcevic and B. M. Rogina. (2007). Quantum random bit generator service for monte carlo and other stochastic simulations. International Conference on Large-Scale Scientific Computing pp. 508515.

Steven, R., J. A. Halderman and A. Finkelstein. (2010). Sketcha, , a CAPTCHA based on line drawings of 3D models. *In*: The Proceedings of the 19th International Conference on World Wide Web. ACM.

SwipAds. (2014). FunCaptcha Anti-Spam CAPTCHA (on-line) https://wordpress.org/plugins/funcaptcha/.

Tam, J., D. Simsa, D. Huggins, L. Von Ahn and M. Blum. (2008). Improving audio CAPTCHAs. *In*: The Proceedings of 4th Symposium on Usability, Privacy and Security (SOUPS 08), Pittsburgh.

Tam, J., J. Simsa, S. Hyde and L. V. Ahn. (2008). Breaking audio captchas. pp. 1625–1632. *In*: Advances in Neural Information Processing Systems.

Tariq Banday, M. and N. A. Shah. (2009). Image flip Captcha. The ISC International Journal of Information Security. Iranian Society of Cryptology 1(2): 105–123.

Von Ahn and Luis. (2008). ReCaptcha: Human-based character recognition via web security measures. Science 321(5895): 1465–1468.

Von Ahn, M. Blum and J. Langford. (2004). Telling humans and computers apart automatically. Communications of the ACM, ACM 47(2): 5660.

Wilkins, J. (2009). Strong Captcha guidelines v1. 2. Retrieved Nov, 10: 8.

Willy, S., CH. Yang-Wai and H. Yu. Zhou. (2012). Ste3d-cap: Stereoscopic 3d Captcha. Cryptology and Network Security. Springer Berlin Heidelberg, pp. 221–240.

Xu, Y., G. Reynaga, S. Chiasson, J. F. Frahm, F. Monrose and V. Oorschot. (2012). Security and usability challenges of moving-object CAPTCHAs: Decoding code-words in motion. *In*: The Proceedings of 21st USENIX Security Symposium, Bellevue, WA, USA.

Yan, J. and A. S. El Ahmad. (2008). A low-cost attack on a microsoft CAPTCHA. pp. 543554. *In*: Proceedings of the 15th ACM Conference on Computer and Communications Security, ACM.

Yang, I. Tzu, Chorng-Shiuh Koong and Chien-Chao Tseng. (2013). Game-based image semantic CAPTCHA on handset devices. Springer, Multimedia Tools and Applications, pp. 1–16.

Yang-Wai, Ch and W. Susilo. (2011). AniCAP: an animated 3d CAPTCHA scheme based on motion parallax. Cryptology and Network Security. Springer, pp. 255–271.

Yang-Wai, Ch. and W. Susilo. (2012). Enhanced STE3D-CAP: a novel 3d CAPTCHA family. Information Security Practice and Experience. Springer, pp. 170–181.

Ye, Q., T. Wei, A. B. Jeng, H. M. Lee and K. Wu. (2013). DDIM-CAPTCH: A novel drag-n-drop interactive masking CAPTCHA against the third party human attacks. Technologies and Applications of Artificial Intelligence (TAAI), Conference on IEEE pp. 158–163.

CHAPTER **7**

Colored Petri Net Model for Blocking Misleading Information Propagation in Online Social Networks

Mohamed Torky[1,3] and *Aboul Ella Hassanien*[2,3,*]

Introduction

Online Social Networks (OSNs) are popular platforms for its users to share a variety of information patterns (Ghali et al. 2012, Abraham et al. 2010). Propagated information patterns may be in the form of good information (i.e., credible and accurate information) or misleading information (i.e., incredible and deceptive information) (Kumar et al. 2014). Misleading information (may be called Rumors or Semantic Attacks (Kumar et al. 2014)) can have serious consequences on decision making of people, firms and governments, since it can create confusion, deception, and mistrust among the receivers (Karlova and Fisher 2013). The propagation of misleading information is especially important in the state of newsworthy information, where new segments of information are shared progressively, often starting off as unverified information and then constitute a rumor pattern. For instance, "Much of the fake news that flooded the internet

---
[1] Faculty of Science, Menofyia University, Egypt.
[2] Faculty of Computers and Information, Cairo University, Egypt.
[3] Scientific Research Group in Egypt (SRGE).
Email: mtorky86@gmail.com; http://www.egyptscience.net
[*] Corresponding author: aboitcairo@gmail.com

during the 2016 election season consisted of written pieces and recorded segments promoting false information or perpetuating conspiracy theories. Some news organizations published reports spotlighting examples of hoaxes, fake news and misinformation on Election Day 2016" (Ordway 2017, Allcott and Gentzkow 2017). The severity of rumors and misleading information is that a single false information from an incredible source can quickly affects thousands of infected users and may cause a huge damage in the financial transactions and stock markets. For example, when crackers hijacked the Associated Presss Twitter account in 2014 and posted a tweet claiming that the White House had been attacked, the Standard & Poor's 500 (or S & P 500) index immediately dropped and lost about 130 billion dollar (Luckerson 2014). Hence, investigating information credibility and detecting rumors is a continuing concern within the field of social network analysis.

Treating such type of information attacks requires firstly recognition of the patterns of misleading information via evaluating information credibility (Gupta and Kumaraguru 2012, Liu 2012, Torky et al. 2016), then working to block its propagation in the social graph. Recently, a considerable literature has grown up around the theme of treating rumors and misleading information problem in social networks (He et al. 2017, Vicario et al. 2016). However, the synthesis of detecting and blocking mechanisms of rumors remains is a major challenge in social network platforms.

This paper attempts to show the feasibility of employing Colored Petri Nets (CPNs) (Jensen 2013) for modelling the proposed solution of detecting and blocking rumors patterns. CPNs is a high-level class of Petri nets that involve additional features than classical class of Petri nets. In CPN, *tokens* appear into several datatype (i.e., set of colors). *Places* used as a repository of tokens in the same colored set. *Transitions* used to perform specific functions based on achieving a *guard expression*. Each *arc* is labeled with an *arc expression*, where the input and output types of the arc expressions must correspond to the type of places the arc is connected to. A transition t is said to be enabled if there are enough tokens in the 'right color' in each input place of t. The right color concept means that there is a consistent substitution of tokens' colors into the corresponding arc expression. Updating the net from one state to another is represented as a set of *marking states*. CPN models must be associated with a *declaration panel* to describe the color set of tokens and the variables used in arc expression. Figure 1 explains an example of enabling-firing process in CPNs. Enabling state of transition t_1 in Fig. 1(a) demonstrates that t_1 has two input places over two arc expressions x and y. The variable x can be substituted with an integer value of token from place P_1, which may be 8 or 3. Variable y can be substituted with an integer value of token from

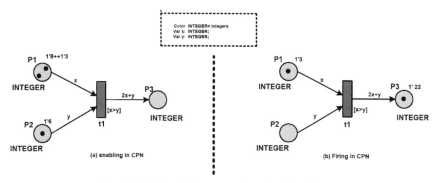

Fig. 1. Enabling-firing process in colored petri nets.

place P_2, which is 6. Transition t_1 is associated with the guard expression $x > y$ that returns boolean value for allowing firing process. Hence, variable x must be substituted with token's value 8 and variable y is substituted with the token value 6. Firing t_1 produces a new token value in an integer type according to the out-arc expression $2x + y$. Hence, place P_3 is marked with one token has a value 22 as depicted in Fig. 1(b).

In this paper, we introduce a novel Colored Petri Net Model (CPNM) for modeling the solution of detecting misleading information-tokens, and blocking its propagation across OSNs. The proposed CPNM is experimentally simulated and evaluated on two datasets consisting of 1003 and 863-newsworthy tweets collected from Twitter social networks. The experimental results demonstrated outperforming of CPNM in detecting rumors tweets compared with other mechanisms in the literature with respect to the metrics of Precision, Recall, Accuracy, and Fale Positive Rate (FPR). In addition, the reachability analysis of the proposed CPNM proved that *detecting*, and *blocking* rumors tokens (i.e., misleading tweets) are reachable states with respect to the firing sequence of tokens.

The rest of this paper can be organized into six sections. Section 2, Related Work, presents the literature review. Section 3, Detecting and Blocking Rumors Model, presents the proposed CPNM approach. Section 4, Reachability Analysis of CPNM, investigates and analyzes the reachability property of CPNM. Section 5, Simulation and Practical Results, presents a simulation and practical results of CPNM functionality. Section 6, Discussion, discusses and interprets the obtained results. Section 7, formulates the general conclusion of this study.

Related Work

Most of the proposed approaches in the literature focused on verifying the content or the sources of shared information to identify misleading information without introducing an appropriate mechanism to block its diffusion over OSNs. Kumar et al. (2014) introduced a cognitive psychology-based approach for detecting misinformation in online social networks. The proposed approach depends on verifying information consistency, information coherency, the credibility of sources and general acceptability of message. A novel ranking approach (Gupta and Kumaraguru 2012, Torky et al. 2016, Cstillo et al. 2011, Morris et al. 2012, Abbasi and Liu 2013) is proposed to evaluate users' and tweets' credibility in Twitter based on some features of tweet-content. Other works depended on verifying the source of misleading information in social networks, such as ranking-based algorithm and optimization-based algorithm (Nguyen et al. 2014), and identifying rumor source-based algorithm (Qazvinian 2011). K-Effectors mechanism (Lappas et al. 2010) is proposed to identify a set of k nodes that control the current activation status of the social network. Nguyen et al. (2013) introduced a mathematical approach to limit viral propagation of misinformation in OSNs. The authors study the family of β_T^I node protector problems for decontaminating misinformation with good information. Budak et al. (2014) conducted an extensive study of the problem of limiting the propagation of misinformation in a social network, but their results proved that this problem is NP-hard and therefore an exact solution is not feasible for large-scale social networks.

Detecting and Blocking Rumors Model

In this section, we propose a novel colored petri net approach for detecting and blocking misleading information propagation instead of the decontamination techniques that are proposed in (Lappas 2010, Nguyen 2013). In our preliminary work in Torky et al. (2016), we introduced a mathematical model for ranking newsworthy tweets based on evaluating and verifying the credibility of information sources with respect to the URL (Unified Resource Locator) feature in the tweet content. We seek to design and develop a more mature mechanism for classifying the shared information as *good information* (i.e., credible information) and *misleading information* (i.e., rumors) before working to block the propagation of detected rumors patterns. In response to this challenge, the proposed CPNM can perform two major functions: detecting rumors-tokens based on proposed credibility evaluation algorithms assigned to the set of transitions and blocking the propagation of detected rumors-tokens. According to the modeling language in colored petri nets, we can represent processing information-credibility as a set of *marking states*: $R(M_0) = \{M_0,$ $M_1, M_2....M_n\}$, such that each marking state M_i is represented as a finite set of *colored places* in the form: $M_i = (P1, P2, P3, ...Pn)$. The shared information patterns can be represented as a set of *colored tokens* in the form $x_i v_i$, such that x_i is the number of occurrences of the token, and v_i is the token-data type (or color set). The algorithms used to evaluate information credibility can be represented as a set of *transitions, t1, t2, t3, ...tn*, and the net change from one state to another state can be represented using a set of arcs labeled with inspections in the form of mathematical expressions that control the firing process. The proposed CPNM is described in Fig. 2. The declaration panel of CPNM net involves nine color sets. The color set for each place of CPNM is depicted in Table 1.

The initial marking M_0 in the proposed CPNM is initialized by allocating place $P1$ with seven multi-set elements that represent the common kinds of information sources (i.e., colored tokens), which are

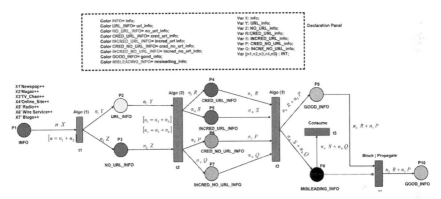

Fig. 2 The proposed colored petri net model (CPNM).

Table 1. Color sets of ten places in the proposed CPNM net.

Places	Color Set	Meaning
P1	INFO	Information
P2	URL_INFO	Information contain URL
P3	NO_URL_INFO	Information doesn't contain URL
P4	CRED_URL_INFO	Credible URL-information
P5	INCRED_URL_INFO	Incredible URL-information
P6	CRED_NO_URL_INFO	Credible No-URL-information
P7	INCRED_NO_URL_INFO	Incredible No-URL-information
P8	GOOD_INFO	Good information
P9	MISLEADING_INFO	Misleading information
P10	GOOD_INFO	Good information

newspapers, magazines, TV channels, online sites, radio, wire services, and blogs. The number of tokens' occurrences for each source element are represented by the variables $x1, x2, x3, x4, x5, x6, x7$ respectively. Firing the transition $t1$ will classify the input tokens (e.g., tweets) according to the arc expression nX into two classes of information in the places $P2$, and $P3$. The produced token in $P2$ is based on the out-arc expression n_1Y and the produced token in $P3$ is based on the out-arc expression n_2Z, where, $P2$ is the repository of URL-tokens, and place $P3$ is the repository of No-URL-tokens. The major functionality of $t1$ is depicted in Algorithm 1.

Firing $t1$ depends on achieving the guard expression $n = n1 + n2$, where, n is the number of all input tokens in the color set *INFO*, $n1$ is the number of tokens in the color set *URL_INFO* and $n2$ is the number of tokens in the color set *NO_URL_INFO*. The firing process of $t2$ is based on two input arc-expressions—n_1Y and n_2Z for evaluating the credibility of all tokens in places $P2$, and $P3$. Firing $t2$ produces four kinds of tokens in places $P4, P5, P6$, and $P7$ with respect to four output arc expressions—n_3R, n_4S, n_5P and n_6Q respectively. The four places represent four levels of information credibility according to the color set of each place. $P4$ is the

---

Algorithm 1 Transition $t1$: Information Classification

1: **Input:** Info $F = \{F_1, F_2, F_3, ..F_n\}$
2: **Output:** URL_Info $UF = \{F_1, F_2, F_3, ..F_{n1}\}$
3: **Output:** No_URL_Info $NUF = \{F_1, F_2, F_3, ..F_{n2}\}$
4: **procedure** INFO-CLASSIFICATION
5:
6:     **for** $F_i \in F$ **do**
7:         **if** *IsTrue*$(F_i.url)$ **then**
8:         $UF = UF \cup F_i$
9:     **else**
10:        $NUF = NUF \cup F_i$
11:    **end if**
12:    **end for**
13: **return** $UF = \{F_1, F_2, F_3, ..F_{n1}\}$
14: **return** $NUF = \{F_1, F_2, F_3, ..F_{n2}\}$
15:    \\Guard Expression Condition
16:    **if** $n == n_1 + n_2$ **then**
17: **return** *True*
18:    **else**
19: **return** *False*
20:    **end if**
21: **end procedure**

repository of all tokens that represent *Credible URL-information*, P5 is the repository of all tokens that represent *Incredible URL-information*, P6 is the repository of all tokens that represent *Credible No-URL-information*, and P7 is the repository of all tokens that represent *Incredible No-URL-information*. In addition, Firing $t2$ depends on achieving two guard expressions: $n1 = n3 + n4$ and $n2 = n5 + n6$ such that, $n3$ is the number of tokens in the color set *CRED_URL_INFO*, $n4$ is the number of tokens in the color set *INCRED_URL_INFO*, $n5$ is the number of tokens in the color set *CRED_NO_URL_INFO*, and $n6$ is the number of tokens in the color set *INCRED_NO_URL_INFO*. Algorithm 2 demonstrates the methodology of evaluating information credibility as an assigned functionality to transition $t2$.

Firing $t3$ according to the two input arc expressions n_3R and n_5P unifies the tokens in P4 with the tokens in P6 and produces the unification result into places P8, according to the output-arc expression $n_3R + n_5P$. The all produced tokens in P8 are in the form good information-tokens. At the same time, firing $t3$ according to the other two input arc expressions n_4S and n_6Q unifies the tokens in P5 with the tokens in P7 and produces the unification result into places P9 according to the output-arc expression $n_4S + n_6Q$. The tokens in P9 are in the form of misleading information-tokens (i.e., rumors). The major functionality of transition $t3$ is depicted in the Algorithm 3. Algorithm 4 explains that enabling or disabling transition $t4$ depends on the *inhibitor arc*. The inhibitor arc disables the transition $t4$ and blocks all tokens in place P9.

Hence, firing $t4$ can be done if all tokens in place P9 have been removed. Hence, firing the transition $t5$ firstly will remove (consume) all rumors tokens in P9. Then transition $t4$ becomes enabled and can fire all tokens (i.e., good information tokens) in place P8 and produces them as output tokens in place P10.

Reachability Analysis of CPNM

Reachability is the most important behavior property to verify token behaviors in petri nets. A marking state M_n is said to be reachable from the initial marking state M_0 if there is a firing sequence $\sigma = t_1, t_2, t_3, ...t_m$ that transforms M_0 to M_n. In other words, if the set of all possible markings reachable from M_0 in the net (CPN, M_0) is denoted by $R(M_0)$, the reachability problem is to prove that the marking state $M_n \in R(M_0)$. With respect to the dynamic behaviour of the proposed CPNM, it is mandatory to prove that detecting rumors token and blocking its propagation (firing) are reachable states from the initial marking M_0. The reachability problem in the proposed CPNM can be formulated as in the following theorem:

Algorithm 2 Transition $t2$: Information Credibility Evaluation

---

1: **Input:** $URL\_Info\ UF = \{F_1, F_2, F_3, ..F_{n1}\}$
2: **Input:**$No\_URL\_Info\ NUF = \{F_1, F_2, F_3, ..F_{n2}\}$
3: **Output:** $Cred\_URL\_Info\ CUF = \{F_1, F_2, F_3, ..F_{n3}\}$
4: **Output:** $Incred\_URL\_Info\ ICUF = \{F_1, F_2, F_3, ..F_{n4}\}$
5: **Output:** $Cred\_No\_URL\_Info\ CNUF = \{F_1, F_2, F_3, ..F_{n5}\}$
6: **Output:** $Incred\_No\_URL\_Info\ ICNUF = \{F_1, F_2, F_3, ..F_{n6}\}$
7: **procedure** CREDIBILITY EVALUATION
8:
9:     **for each** $F_i \in UF$ **do**
10:         Score(F_i)=BM25F$(F_i.url)$ [5]
11:     **end for**
12:     $Cred_{Threshold1} = \sum_{i=1}^{i=n1} \frac{Score(F_i)}{n1}$
13:     **for each** $F_i \in UF$ **do**
14:         **if** $Score(F_i) \geq Cred_{Threshold1}$ **then**
15:             $CUF = CUF \cup F_i$
16:         **else**
17:             $ICUF = ICUF \cup F_i$
18:         **end if**
19:     **end for**
20:     **for each** $F_i \in NUF$ **do**
21:         ER(F_i)=$\frac{RE+RT}{FL} \times 100$
22:     **end for**
23:     $Cred_{Threshold2} = \sum_{i=1}^{i=n2} \frac{ER(F_i)}{n2}$
24:     **for each** $F_i \in NUF$ **do**
25:         **if** $ER(F_i) \geq Cred_{Threshold2}$ **then**
26:             $CNUF = CNUF \cup F_i$
27:         **else**
28:             $ICNUF = ICNUF \cup F_i$
29:         **end if**
30:     **end for**
31: **return** $CUF = \{F_1, F_2, F_3, ..F_{n3}\}$
32: **return** $ICUF = \{F_1, F_2, F_3, ..F_{n4}\}$
33: **return** $CNUF = \{F_1, F_2, F_3, ..F_{n5}\}$
34: **return** $ICNUF = \{F_1, F_2, F_3, ..F_{n6}\}$
35:     \\Guard Expression Condition
36:     **if** $(n1 == n_3 + n_4\ \&\&\ n2 == n_5 + n_6)$ **then**
37: **return** $True$
38:     **else**
39: **return** $False$
40:     **end if**
41: **end procedure**

Algorithm 3 Transition $t3$: Good/Misleading Information detection

---

1: **Input:** $Cred\_URL\_Info \ CUF = \{F_1, F_2, F_3, ..F_{n3}\}$
2: **Input:** $Incred\_URL\_Info \ ICUF = \{F_1, F_2, F_3, ..F_{n4}\}$
3: **Input:** $Cred\_No\_URL\_Info \ CNUF = \{F_1, F_2, F_3, ..F_{n5}\}$
4: **Input:** $Incred\_No\_URL\_Info \ ICNUF = \{F_1, F_2, F_3, ..F_{n6}\}$
5: **Output:** $Good\_Info \ GF = \{F_1, F_2, F_3, ..F_{n7}\}$
6: **Output:** $Misleading\_Info \ MF = \{F_1, F_2, F_3, ..F_{n8}\}$
7: **procedure** GOOD/ MISLEADING-INFO-DETECTIONS
8:
9:     $GF = CUF \cup CNUF$
10:     $MF = ICUF \cup ICNUF$
11: **return** $GF = \{F_1, F_2, F_3, ..F_{n7}\}$
12: **return** $MF = \{F_1, F_2, F_3, ..F_{n8}\}$
13: **end procedure**

Algorithm 4 Transition $t4$: Propagating/Blocking Information

---

1: **Input:** $Good\_Info \ GF = \{F_1, F_2, F_3, ..F_{n7}\}$
2: **Input:** $Misleading\_Info \ MF = \{F_1, F_2, F_3, ..F_{n8}\}$
3: **procedure** PROPAGATE/ BLOCK INFORMATION
4:
5:     **if** $MF = \phi$ **then**
6:         $CPN_{Output} \leftarrow Propagate(GF)$
7:     **else**
8:         $CPN_{Output} \leftarrow Block(MF)$
9:     **end if**
10: **end procedure**

Theorem: Given the proposed CPNM net ($CPNM$, M_0), with an initial marking M_0 such that $M_0 = (n, 0, 0, 0, 0, 0, 0, 0, 0, 0)$ and knowing the variables $n, n1, n2, n3, n4, n5, n6$, where $n = n1 + n2$, $n1 = n3 + n4$, and $n2 = n5 + n6$, the following points can be proved:

1. $M_3 \in R(M_0)$, such that $M_3 = (0, 0, 0, 0, 0, 0, 0, n3 + n5, n4 + n6, 0)$
2. $M_4 \in R(M_0)$, such that $M_4 = (0, 0, 0, 0, 0, 0, 0, n3 + n5, 0, 0)$
3. $M_5 \in R(M_0)$, such that $M_5 = (0, 0, 0, 0, 0, 0, 0, 0, 0, n3 + n5)$

Proof

Supposing that the initial marking of a given net ($CPNM$, M_0) is:

$$M_0 = (n, 0, 0, 0, 0, 0, 0, 0, 0, 0) \tag{1}$$

So,

$$R(M_0) = \{M_0\} \tag{2}$$

if $\sigma = t_1$, Then,

$$M_0 \Rightarrow M_1 = (0, n1, n2, 0, 0, 0, 0, 0, 0, 0) \tag{3}$$

So,

$$R(M_0) = \{M_0, M_1\} \tag{4}$$

if $\sigma = t_1, t_2$, Then,

$$M_1 \Rightarrow M_2 = (0, 0, 0, n3, n4, n5, n6, 0, 0, 0) \tag{5}$$

So,

$$R(M_0) = \{M_0, M_1, M_2\} \tag{6}$$

if $\sigma = t_1, t_2, t_3$, Then,

$$M_2 \Rightarrow M_3 = (0, 0, 0, 0, 0, 0, 0, n3 + n5, n4 + n6, 0) \tag{7}$$

So,

$$R(M_0) = \{M_0, M_1, M_2, M_3\} \tag{8}$$

Hence,

$$M_3 \in R(M_0) \tag{9}$$

Equation 9 proves that the marking state M_3 is a reachable state from the initial marking M_0. In this state, the n colored tokens (i.e., information patterns) are classified into two classes—the first class involves $n3 + n5$ tokens in place P_8 as good information patterns; the second class involves $n4 + n6$ tokens in place P_9 as rumor information patterns according to the firing sequence $\sigma = t_1, t_2, t_3$.

if $\sigma = t_1, t_2, t_3, t_s$ Then,

$$M_3 \Rightarrow M_4 = (0, 0, 0, 0, 0, 0, 0, n3 + n5, 0, 0) \tag{10}$$

So,

$$R(M_0) = \{M_0, M_1, M_2, M_3, M_4\} \tag{11}$$

Hence,

$$M_4 \in R(M_0) \tag{12}$$

Equation 12 proves that the marking state M_4 is a reachable state from the initial marking M_0. In this state, the $n4 + n6$ tokens in place P_9 that represent rumors information patterns are blocked and removed from P_9 according to the firing sequence $\sigma = t_1, t_2, t_3, t_5$.

if $\sigma = t_1, t_2, t_3, t_5, t_4$, Then,

$$M_4 \Rightarrow M_5 = (0, 0, 0, 0, 0, 0, 0, 0, 0, n3 + n5) \tag{13}$$

So,

$$R(M_0) = \{M_0, M_1, M_2, M_3, M_4, M_5\} \tag{14}$$

Hence,

$$M_5 \in R(M_0) \tag{15}$$

Equation 15 proves that the marking state M_5 is a reachable state from the initial marking M_0. In this state, the $n3 + n5$ tokens that represent the good information patterns are produced as a final output token in place P_{10}. In general, in any $(CPNM, M_0)$, M_5 is the final state in which $n3 + n5$ tokens that represent good information patterns are the final output of the token-ring process, according to the firing sequence $\sigma = t_1, t_2, t_3, t_5, t_4$.

It is clear that the proposed net $(CPNM, M_0)$ is a live net, such that all transitions can be fired at least one time according to the firing sequence $\sigma = t_1, t_2, t_3, t_5, t_4$. Hence, the five transitions are L1-live. In addition, the proposed $(CPNM, M_0)$ isn't reversible, where, for each marking state M_n in $R(M_0)$, the initial marking M_0 isn't reachable from M_n. This means that the proposed $(CPNM, M_0)$ is an input-output net. The maximum number of times that one transition t_x is firing while the other transition t_y isn't firing is only one. This observation clarifies that any pair of transitions in the proposed $(CPNM, M_0)$ are in (1-faire) relation.

Example: Suppose we have a net $(CPNM, M_0)$ as in Fig. 3, where the initial marking state is $M_0 = (7, 0, 0, 0, 0, 0, 0, 0, 0, 0)$. This means that place P_1 is initialized with seven tokens and all places from P_2 to P_{10} have no tokens with respect to the color set of each place, as depicted previously in Table 2. Secondly, assume that the enabled bindings of variables n, $n1$, $n2$, $n3$, $n4$, $n5$, $n6$ have values 7, 4, 3, 1, 3, 2, 1 respectively. The tokens distribution in $(CPNM, M_0)$ can be executed according to the firing sequence $\sigma = t_1, t_2, t_3, t_5, t_4$ as follows:

Firing t_1 produces four tokens in place P_2 and three tokens in place P_3 according to the color set of each place as depicted previously in Table 2. Then, the initial marking state $M_0 = (7, 0, 0, 0, 0, 0, 0, 0, 0, 0)$ is changed to $M_1 = (0, 4, 3, 0, 0, 0, 0, 0, 0, 0)$. Figure 4 presents the enabling-firing processes of the transition t_1. Firing t_1 substitutes seven tokens in place P_1 with the expression $7X$ and produces four tokens in place P_2 according to its color-set, where, the four tokens are substituted with the expression $4Y$. Moreover, it produces three tokens in place P_3 according to its color-set, where, the three tokens are substituted with the expression $3Z$.

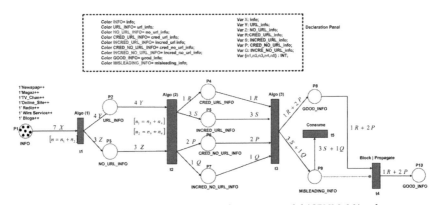

Fig. 3. An example: the proposed colored petri net model (*CPNM*, M_0), where, $M_0 = (7, 0, 0, 0, 0, 0, 0, 0, 0, 0)$.

Table 2. Number of tweets according to its sources in the four datasets.

Source	#Tweets in #ISIS	#Tweets in #CharlieHebdo
Newspapers	281	116
Magazines	149	87
TV Channels	132	101
Radios	99	76
Online Sites	114	174
Wire Services	109	92
Blogs	119	217
Sum	**1003**	**863**
Initial Marking M_0	$M_0 = (1003, 0, 0, 0, 0, 0, 0, 0, 0, 0)$	$M_0 = (863, 0, 0, 0, 0, 0, 0, 0, 0, 0)$

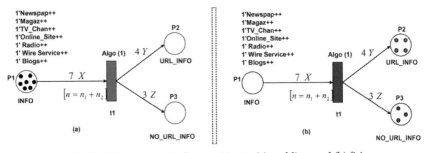

Fig. 4. The firing process in the transition t_1: (a) enabling, and (b) firing.

With respect to the color-set of each output places of transition t_2, firing t_2 produces one token in place P_4 according to the expression $1R$, three tokens in place P_5 according to the expression $3S$, two tokens in place P_6

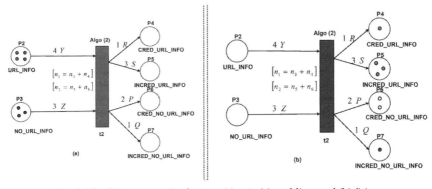

Fig. 5. The firing process in the transition t_2: (a) enabling, and (b) firing.

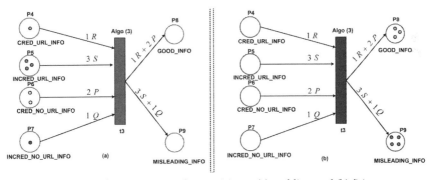

Fig. 6. The firing process in the transition t_3: (a) enabling, and (b) firing.

according to the expression $2P$, and one token in place P_7 according to the expression $1Q$. Then, the marking state is changed to $M_2 = (0, 0, 0, 1, 3, 2, 1, 0, 0, 0)$. Figure 5 presents the enabling-firing process of the transition t_2.

With respect to the color-set of each output places of transition t_3, firing t_3 produces three tokens in place P_8 according to the expression $1R + 2P$, and produces four tokens in place P_9 according to the expression $3S + 1Q$.

This means that, the three tokens detected in place P_8 represent good information patterns and the four tokens detected in place P_9 represent misleading information patterns (i.e., rumors). Hence, the marking state is changed to $M_3 = (0, 0, 0, 0, 0, 0, 0, 3, 4, 0)$. Figure 6 presents the enabling-firing processes of transition t_3.

The inhibitor arc (i.e., the dashed line terminated with circle) from P_9 to transition t_4 disables t_4 as long as place P_9 contains any token, but, t_5 is enabled and can fire for consuming the four tokens in P_9 with respect to the expression $3S + 1Q$. Firing t_5 changes the net to the marking state $M_4 = (0, 0, 0, 0, 0, 0, 0, 3, 0, 0)$. In this marking state of $(CPNM, M_0)$, all rumors tokens in P_9 are blocked and then removed from the net. Figure 7 presents the enabling-firing process of the transition t_5.

Firing transition t_4 transfers three tokens from P_8 to P_{10} according to expression $1R + 2P$ as a final output in the firing sequence. The marking state is changed to $M_5 = (0, 0, 0, 0, 0, 0, 0, 0, 0, 3)$. Figure 8 presents the enabling-firing process of t_4. The set of all marking states is $R(M_0) = \{M_1, M_2, M_3, M_4, M_5\}$ according to the firing sequence $\sigma = t_1, t_2, t_3, t_5, t_4$. The reachability analysis demonstrated that M_3, M_4 and M_5 are reachable states with respect to tokens firing process. $M_3 = (0, 0, 0, 0, 0, 0, 0, 3, 4, 0)$ represents detecting three tokens as good information patterns and four tokens as rumors patterns.

Moreover, $M_4 = (0, 0, 0, 0, 0, 0, 0, 3, 0, 0)$ represents blocking and removing four tokens as rumors patterns, and $M_5 = (0, 0, 0, 0, 0, 0, 0, 0, 0, 3)$ represents producing three tokens in place P_{10} as good information patterns represent the final output. Figure 9 provides all marking states

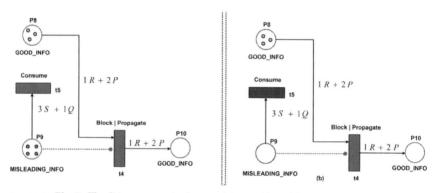

Fig. 7. The firing process in the transition t_5: (a) enabling, and (b) firing.

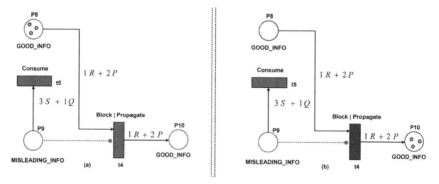

Fig. 8. The firing process in the transition t_4: (a) enabling, and (b) firing.

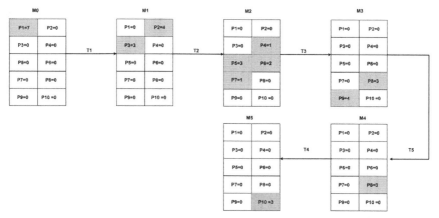

Fig. 9. Reachability graph model of CPNM with respect to the initial marking $M_0 = (7, 0, 0, 0, 0, 0, 0, 0, 0, 0)$.

as a reachability graph model according to the initial marking state $M_0 = (7, 0, 0, 0, 0, 0, 0, 0, 0, 0)$.

Simulation and Practical Results

For investigating the realism and effectiveness of the proposed CPNM mechanism, we used the *CPN simulation tool* (Simulator 2011) for simulating the functionality of the proposed model on two datasets of tweets collected from Twitter social network. The two datasets are described as two trending topics (i.e., #ISIS and #CharlieHebdo) involving several newsworthy tweets shared by multiples sources of information. We used the *Twitter R library* (Gentry 2011) as a software tool for collecting tweets of each trending topic. We ignored some tweets in each dataset since these tweets either haven't correct URL or haven't the screen name of the source of information. The general description of the two datasets and the initial marking states of the proposed CPNM are described in Table 2.

We have categorized the collected tweets in each dataset according to its information sources (i.e., newspapers, magazines, TV channels, radios, wire service, and blogs). We implemented the functionality of Algorithms 1, 2, 3, 4 in each dataset for detecting rumors patterns in each one; then, we simulated the proposed CPNM in each dataset for verifying the detection and blocking of rumors patterns. In the first round of simulation, we initialized the proposed CPNM as $M_0 = (1003, 0, 0, 0, 0, 0, 0, 0, 0, 0)$, where place P_1 is initialized with 1003 tokens (i.e., tweets in #ISIS dataset). In the second round of simulation, we initialized the proposed CPNM as $M_0 = (863, 0, 0, 0, 0, 0, 0, 0, 0, 0)$, where place P_1 is initialized with 863 tokens (i.e., tweets in #CharlieHebdo dataset).

Dataset 1 (#ISIS)

#ISIS a hot trending topic in Twitter, has been created currently with appearance of the Islamic State of Iraq and al-Sham (ISIS). Variety of newsworthy tweets have been propagated around this trendy topic in Twitter social network. Detecting and blocking rumors in #ISIS dataset has been simulated on 1003 tweets represented as a set of colored tokens used to initialize the proposed net ($CPNM$, M_0) as M_0 = (1003, 0, 0, 0, 0, 0, 0, 0, 0, 0). Table 3 summarizes the token distributions in the ten places according to the firing sequence $\sigma = t_1, t_2, t_3, t_5, t_4$. The simulation results demonstrated that 679-tokens have been detected, then blocked as rumors tokens (i.e., misleading information). 324-tokens have been detected and produced as credible-tokens (i.e., good information).

Figure 10 presents the percentages of tokens according to the colorset of each place in the set of ten places P_1, P_2, ...P_{10}. Figure 11 presents a pie plot, which provides the percentages of detected good tokens (i.e., credible

Table 3. Tokens-distribution according to the firing sequence $\sigma = t1, t2, t3, t5, t4$ in ($CPNM$, M_0) when simulated on #ISIS dataset.

Transitions	P1	P2	P3	P4	P5	P6	P7	P8	P9	P10
t1	0	719	284	0	0	0	0	0	0	0
t2	0	0	0	137	582	187	97	0	0	0
t3	0	0	0	0	0	0	0	324	679	0
t5	0	0	0	0	0	0	0	324	0	0
t4	0	0	0	0	0	0	0	0	0	324

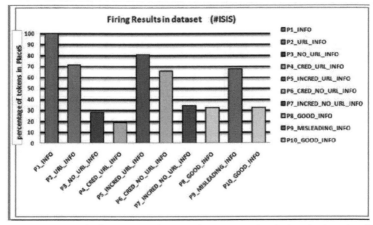

Fig. 10. Percentages of tokens in ten places with respect to simulating the functionality of ($CPNM$, M_0) in #ISIS dataset.

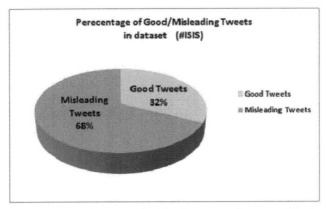

Fig. 11. Percentages of good tweets and misleading tweets (i.e., rumors) in dataset #ISIS.

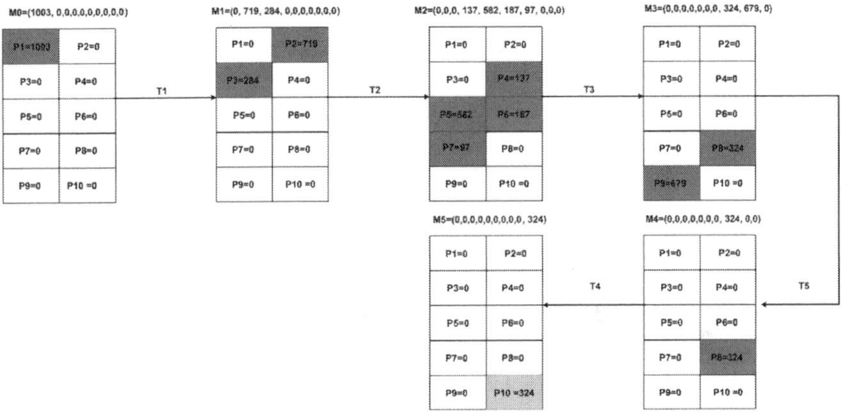

Fig. 12. Reachability graph model when simulating $(CPNM, M_0)$ net on dataset (#ISIS).

tweets) and rumors tokens (i.e., misleading tweets) in #ISIS dataset. Figure 12 provides the reachability graph of all marking states when simulating $(CPNM, M_0)$ net on dataset (#ISIS).

Dataset 2 (#CharlieHebdo)

#CharlieHebdo is another Bot trending topic propagated in Twitter. A lot of newsworthy tweets have been tweeted around the issue of insulting the Prophet Muhammad published in CharlieHebdo magazine. Detecting and blocking rumors tweets in #CharlieHebdo dataset has been simulated on 863 tweets represented as set of colored tokens used to initialize the proposed net $(CPNM, M_0)$ as $M_0 = (863, 0, 0, 0, 0, 0, 0, 0, 0, 0)$. Table 4 provides the token distributions in the ten places according to the firing

Table 4. Tokens-distribution according to the firing sequence $\sigma = t1, t2, t3, t5, t4$ in (*CPNM*, M_0) when simulated on #CharlieHebdo dataset.

Transitions	P1	P2	P3	P4	P5	P6	P7	P8	P9	P10
t1	0	347	516	0	0	0	0	0	0	0
t2	0	0	0	263	84	294	222	0	0	0
t3	0	0	0	0	0	0	0	557	306	0
t5	0	0	0	0	0	0	0	557	0	0
t4	0	0	0	0	0	0	0	0	0	557

sequence $\sigma = t_1, t_2, t_3, t_5, t_4$. The simulation results clarified that 306-tokens have been detected; then blocked as rumors tokens. 557-tokens have been detected before being produced as credible-tokens.

Figure 13 presents the percentages of tokens according to the colorset of each place and Fig. 14 presents a pie plot, which provides the percentages of detected good tokens (i.e., credible tweets) and rumors tokens (i.e., misleading tweets) in this dataset. Figure 15 provides the reachability graph of all marking states when simulating (*CPNM*, M_0) net on dataset (#CharlieHebdo).

According to the well-known matrix of pattern recognition and information retrieval with binary classification. Table 5 summarizes the calculations of evaluating the performance of the proposed CPNM approach in detecting misleading information tokens in the two datasets—*#ISIS, #CharlieHebdo*.

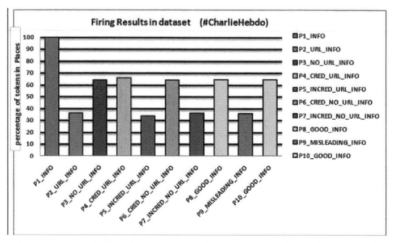

Fig. 13. Percentages of tokens in ten places with respect to simulating the functionality of (*CPNM*, M_0) in dataset (#CharlieHebdo).

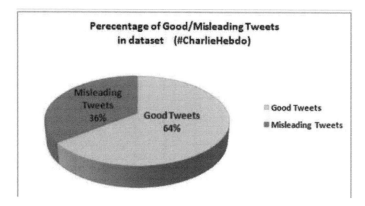

Fig. 14. Percentages of good tweets and misleading tweets (i.e., rumors) in dataset #CharlieHebdo.

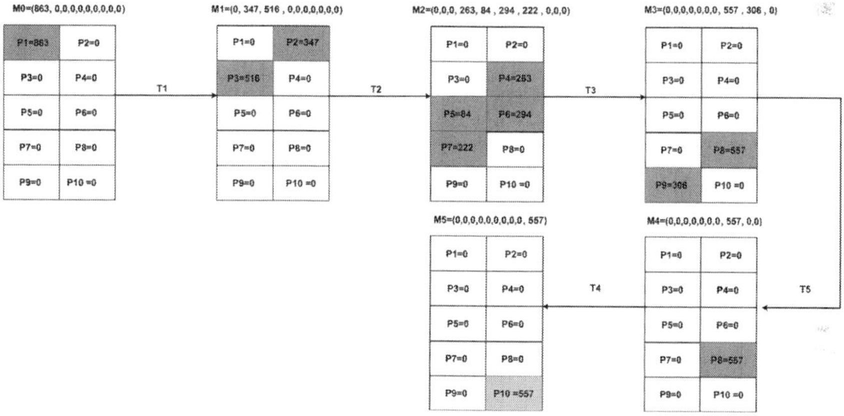

Fig. 15. Reachability graph model when simulating $(CPNM, M_0)$ net on dataset 2 (*#CharlieHebdo*).

Figure 16 compares performance of the proposed CPNM in the two handled datasets in terns of *Precision, Recall, Specificity, Negative Predictive Value (NPV)* and *Accuracy*. Moreover, Fig. 17 compares the error rates, FPR and FDR, of the proposed CPNM.

Discussion

This study set out with the aim of assessing the importance of detecting rumors patterns and blocking its propagation in OSNs systems. Previous studies have noted the importance of detecting rumors, however, very

Table 5. Evaluating the performance of the proposed CPNM approach in the two datasets.

Metric	Formula	(#ISIS)	(#CharlieHebdo)	AVG
True Positives (TP)	Correct Detection	585T	279T	432
False Positives (FP)	Incorrect Detection	94T	27T	61T
True Negatives (TN)	Correct Rejection	271T	495T	383T
False Negatives (FN)	Incorrect Rejection	53T	62T	58T
Condition Positives (P)	P = TP + FN	638T	341T	490T
Condition Negatives (N)	N = FP + TN	365T	522T	44T
Precision (PPV)	$PPV = \dfrac{TP}{TP + FP}$	0.86	0.91	0.89
Recall (TPR)	$TPR = \dfrac{TP}{TP + FN}$	0.91	0.82	0.87
Specificity (TNR)	$TNR = \dfrac{TN}{FP + TN}$	0.74	0.95	0.85
Negative Predictive Value (NPV)	$NPV = \dfrac{TN}{TN + FN}$	0.83	0.89	0.86
False Positive Rate (FPR)	$FPR = \dfrac{FP}{FP + TN}$	0.26	0.05	0.16
False Discovery Rate (FDR)	$FDR = \dfrac{FP}{FP + TP}$	0.14	0.09	0.12
False Negative Rate (FNR)	$FNR = \dfrac{FN}{FN + TP}$	0.08	0.18	0.13
Accuracy	$Acc = \dfrac{TP + TN}{P + N}$	**0.85**	**0.90**	**0.88**

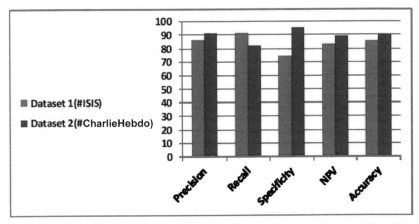

Fig. 16. CPNM-performance evaluation in the two handled datasets in terms of precision, recall, specificity, NPV, and accuracy.

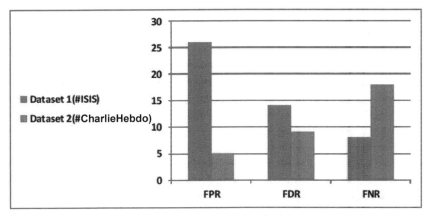

Fig. 17. CPNM-performance evaluation in the two handled datasets in terms of FPR, FDR, and FNR.

little was found in the literature on blocking rumors propagation over OSNs platforms (Lappas 2010, Nguyen 2014, Budak 2011). The present study introduces a *colored petri net model (CPNM)* for solving this problem in a simulated fashion. One interesting finding is that *detecting* and *blocking* rumors are reachable markings states with respect to verifying the reachability problem in the proposed net *(CPNM, M_0)*. The results of the practical simulation study indicated that the correct detection of rumors tokens in the two handled datasets—#*ISIS*, and #*CharlieHebdo* are 585-tweets, and 279-tweets respectively. However, the results showed that the proposed CPNM failed to recognize some rumors patterns in the two datasets. The simulation study showed that the correct rejection of rumors tokens in the two datasets are 271-tweets, and 495-tweets respectively. The most interesting finding was that the proposed CPNM achieved an average precision rate of 89 per cent, recall rate 87 per cent, and accuracy rate 88 per cent in detecting rumors tokens. This finding means that the proposed approach achieved a satisfactory level of exactness, completeness and reliability in detecting rumors patterns with respect to the practical simulation study.

Another important finding is that detecting rumors patterns based on verifying the credibility of information sources by investigating the URL features is in accord with recent studies (Kumar 2014, Qazvinian 2011). Figure 18 provides the comparison results with other mechanisms in terms of detecting rumors patterns in Twitter based on different features that may affect the credibility of tweet (Qazvinian 2011).

Another interesting finding is that reachability graph models depicted previously in Fig. 12 and Fig. 15 demonstrated that detecting and blocking rumors are reachable states when the *(CPNM, M_0)* has been simulated

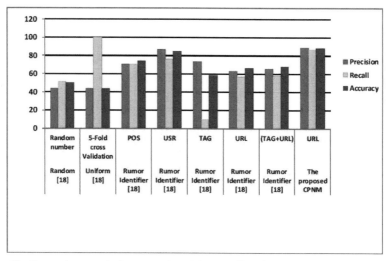

Fig. 18. Comparison results in terms of precision, recall, and accuracy in detecting rumors patterns based on different features of tweets.

on the two handled datasets. These findings may help us to understand the functionality of detecting and blocking rumors patterns in a unified system. However, more extensive credibility evaluation based on other features of published information is needed for recognizing rumors patterns with more accuracy.

Conclusion

In this study, a colored petri net model ($CPNM$, M_0) is introduced for simulating detection and blocking rumors and misleading information in social networks. Novel algorithms have been proposed for investigating information credibility based on the URL feature. The proposed model is verified against reachability and simulated on four datasets as trending topics in Twitter social networks. The reachability verification results proved that detecting rumor tokens and blocking its firing are reachable states with respect to the firing sequence $\sigma = t_1, t_2, t_3, t_5, t_4$ and the set of marking states $R(M_0) = \{M_0, M_1, M_2, M_3, M_4, M_5\}$. The marking state M_3 represents detecting rumors tokens, M_4 represents blocking its firing and M_5 represents producing credible tokens in the final marking state. The performance evaluation disclosed that the proposed CPNM approach achieved a competitive level of exactness (89 per cent), completeness (87 per cent), accuracy (88 per cent), and low false positive rates (FPR) (16 per cent) in detecting rumors-tokens compared with other mechanisms in the literature. Overall, this study strengthens the idea of possibility of detecting and blocking rumors patterns for propagation in OSNs. More research using more controlled trials is needed to investigate the effect of other information features on credibility evaluation for detecting rumors and misleading information with more accuracy.

References

Abbasi, M.A. and H. Liu. (2013). Measuring user credibility in social media. pp. 441–448. *In*: International Conference on Social Computing, Behavioral-Cultural Modeling, and Prediction, Springer Berlin Heidelberg.

Abraham, A., A. E. Hassanien and V. Snsel. (2010). Computational social network analysis: trends, tools and research advances. Springer, Science and Business Media.

Allcott, H. and M. Gentzkow. (2017). Social media and fake news in the 2016 election (No. w23089). National Bureau of Economic Research.

Budak, C., D. Agrawal and A. El Abbadi. (2011). Limiting the spread of misinformation in social networks. pp. 665–674. *In*: Proceedings of the 20th International Conference on World Wide Web ACM, Hyderabad, India.

Cstillo, C., M. Mendoza and B. Poblete. (2011). Information credibility on Twitter. WWW 2011 International Conference, Information Credibility, March 28 April 1, Hyderabad, India.

Del, M., Vicario, A. Bessi, F. Zollo, F. Petroni, A. Scala, G. Caldarelli and W. Quat-trociocchi. (2016). The spreading of misinformation online. Proceedings of the National Academy of Sciences 113(3): 554–559.

Gentry, J. 'Package Twitter R'. (2011). Available (online): http://cran.rproject.org/web/packages/twitterR/twitterR.pdf.

Ghali, N., M. Panda, A. E. Hassanien, A. Abraham and V. Snasel. (2012). Social networks analysis: tools, measures and visualization. pp. 3–23. *In*: Computational Social Networks. Springer London.

Gupta, A. and P. Kumaraguru. (2012). Credibility ranking of tweets during high impact events. Proceedings of the 1st Workshop on Privacy and Security in Online Social Media (PSOSM '12), Article No. 2 ACM New York, NY, USA.

He, Z., Z. Cai, J. Yu, X. Wang, Y. Sun and Y. Li. (2017). Cost-efficient strategies for restraining rumor spreading in mobile social networks. IEEE Transactions on Vehicular Technology 66(3): 2789–2800.

Jensen, K. (2013). Coloured petri nets: basic concepts, analysis methods and practical use. Springer Science & Business Media (Vol. 1).

Karlova, N. A. and K. E. Fisher. (2013). Plz RT: A social diffusion model of misinformation and disinformation for understanding human information behavior. Inform. Res. 18(1): 1–17.

Kumar, K. P. and G. Geethakumari. (2014). Detecting misinformation in online social networks using cognitive psychology. Human-Centric Computing and Information Science, Springer, 4(1): 1–22.

Lappas, T., E. Terzi, D. Gunopulos and H. Mannila. (2010). Finding effectors in social networks. pp. 1059–1068. *In*: Proceedings of the 16th ACM SIGKDD International Conference on Knowledge Discovery and Data Mining, ACM, Washington DC, DC, USA.

Liu, B. (2012). Sentiment analysis and opinion mining. Synthesis lectures on human language technologies. Morgan and Claypool 5(1): 1–167.

Luckerson, V. (2014). Fear, Misinformation, and Social Media Complicate Ebola Fight, http://time.com /3479254/ebola-social-media/ (Access Date: 26/3/2017).

Morris, M. R., S. Counts, A. Roseway, A. Ho and J. Schwarz. (2012). Tweeting is Believing? Understanding Microblog Credibility Perceptions. CSCW 2012 Conference, ACM, 11–15 February, Seattle, Washington, USA.

Nguyen, D. T., N. P. Nguyen and M. T. Thai. (2012). Sources of misinformation in online social networks: who to suspect? pp. 1–6. *In*: The Proceedings of the International Military Communications Conference (2012-MILCOM), IEEE.

Nguyen, N. P., G. Yan and M. T. Thai. (2013). Analysis of misinformation containment in online social networks. Computer Networks, Elsevier 57(10): 2133–2146.

Ordway, D. M. (2017). Fake news and the spread of misinformation. https://journalistsresource.org/studies/society/internet/fake-news-conspiracy-theories-journalism-research (Access Date: 23/5/2017).

Qazvinian, V., E. Rosengren, D. R. Radev and Q. Mei. (2011). Rumor has it: Identifying misinformation in microblogs. pp. 1589–1599. *In*: Proceedings of the Conference on Empirical Methods in Natural Language Processing. Association for Computational Linguistics, EMNLP 2011, Edinburgh, UK.

Simulator functions CPN Tools Homepage (2017) http://cpntools.org/documentation/tasks/simulation/simulator-functions.

Torky, M., R. Babarse, R. Ibrahim, A. E. Hassanien, G. Schaefer and Sh. Y. Zhu. (2016). Credibility investigation of newsworthy tweets using a visualising petri net model. IEEE International Conference on Systems, Man, and Cybernetics (SMC) October 9–12, Budapest, Hungary.

CHAPTER 8

The Conceptual and Architectural Design of Genetic-Based Optimal Pattern Warehousing System

Vishakha Agarwal,[1] *Akhilesh Tiwari,*[1,*] *R.K. Gupta*[1] and
Uday Pratap Singh[2]

Introduction

Nowadays, every real world organization is digitized. From large government organizations, forensic investigative departments to private industries, each of these bodies are maintaining their departmental and customer information over the digital media. Since, the amount of this information is increasing rapidly, there is a need for a proper management system to handle this data effectively and provide easy access to information to the end-user.

During research, in order to simplify the process of analysing the data, the concept of patterns and pattern warehouse as represented by Agarwal et al. 2018 has been evolved, which states that pattern warehouse is a kind of repository which stores the patterns and eases the task of business analyst to extract valuable information out of the data.

In order to achieve this objective, the author in this chapter, introduces the conceptual and architectural design for optimal pattern warehousing

[1] Department of CSE & IT, Madhav Institute of Technology and Science, Gwalior 474005 (M.P.), India.
[2] Department of Applied Mathematics, Madhav Institute of Technology and Science, Gwalior 474005 (M.P.), India.
Emails: agarwal.vishakhacse@gmail.com; iiitmrkg@gmail.com; usinghiitg@gmail.com
* Corresponding author: atiwari.mits@gmail.com

system and also proposes a new repository, called the Optimal Pattern Warehouse (OPWH). The architectural design of Optimal Pattern Warehousing System (OPWS) incorporates several necessary components within its architecture and a genetic-based pattern mining algorithm, which efficiently mines the optimal patterns, out of the pattern warehouse. This architecture deals with several issues related to the pattern management problem, namely pattern optimality, representation, access and storage, etc.

The next sections in this chapter, reveal the key points that show the difference between data warehouse and pattern warehouse, motivation, leading to the creation of optimal pattern warehousing system, its architectural aspect, the algorithm and implementation aspect and analysis of generated patterns along with illustrations.

Data Warehouse vs Pattern Warehouse

Until now, the significance of pattern warehouse has been discussed. Table 1 shows the comparative analysis between the data warehouse and the pattern warehouse.

The mining engine that works upon the data warehouse is OLAP (OnLine Analytical Processing) while in the case of Pattern Warehouse, it

Table 1. Data Warehouse Vs Pattern Warehouse.

Feature	OLAP/Data Warehouse	OLOP/Pattern Warehouse
Characteristics	Informational processing	Analytical and optimization processing
Repository type	Database	Pattern base
The repository is the output of the operation	Data collection	Data mining
Query language	DMQL (Data Mining Query Language)	PMQL (Pattern Manipulation Query Language)
Data	Historical	Condensed or summarized
Source of repository	Discrete relational databases and flat files	Data warehouse
No. of records accessed	Millions	Hundreds
DB size	100 GB to TB	100 MB to GB
No. of users	Hundreds	Few (only high officials)
View	Multidimensional	In form of relationship between data, i.e., patterns
Subset	Data Mart	Pattern Mart
Classification attribute of subsets	Department level	Type of patterns
Processing speed	Slow	Fast
State of patterns	Non-volatile	Volatile

is OLOP (OnLine Optimization Processing). It also summarizes the major distinguishing features of OLAP and OLOP working.

Background

Various contributions have been made in this concept since the introduction of the term pattern. Here, all those contributions are categorized into various aspects of pattern warehousing system which are discussed as below:

Architectural Aspect

The architectural aspect of a pattern warehouse consists of a management system which deals with the creation, storage and management of patterns within the pattern warehouse. Following are some contributions of researchers towards building the architecture of management system for pattern warehousing.

Bartolini et al. (2003), first coined the term pattern and introduced Pattern Base Management System (PBMS) for handling patterns. The author also designed the basic architecture of the PBMS which consisted of three layers, namely pattern layer which is populated with patterns; type layer which contains the patterns of specific type or similar characteristics and lastly, the class layer which holds those patterns that are semantically alike.

Later, Catania et al. (2005) proposed PSYCHO: a prototype system for pattern management. This model relied over the management system as introduced by Bartolini et al. 2003. This pattern management system architecture also had three layers, namely physical layer, middle layer and external layer. The physical layer consists of data sources, pattern base and PBMS while middle layer contains PBMS engine, PML interpreter (pattern manipulation language), PDL interpreter (pattern definition language), formula handler and query processor. The uppermost external layer provides the GUI for pattern analysis by the user.

Kostsifakos et al. (2008) developed a tool called pattern-miner which interfaces the pattern base, data mining module, meta-mining module and pattern comparison module with the user and the user gets the pattern querying results on interacting with the pattern-miner.

Recently, Tiwari and Thakur (2014a), described a pattern warehouse management system (PWMS) which provides tools and techniques for pattern manipulation. The architecture of PWMS consists of four layers— the bottommost layer forms the raw data layer that contains data sources and discrete flat files. Above this layer is data mining engine layer which contains tools that create and update patterns. Above this layer, is middle layer which forms PBMS layer that is further sub-divided into type tier and pattern tier. Finally, the uppermost layer is application layer which represents end-users and applications.

Structural Aspect

The structural aspect of the pattern warehouse includes the schema of the pattern warehouse, the data structures required to store different types of patterns and the syntactic and semantic constraints which are applicable over the pattern warehouse architecture. Various contributions have been given towards the structural aspect of the pattern warehouse. Some of them are discussed below.

Firstly, Rizzi (2004), worked upon the issue of logical modelling of pattern-bases and put an effort to statically model the patterns with the help of Unified Modelling Language (UML). In this way, the author mapped the concept of management and definition of patterns with the object-oriented concepts. The author believes that conceptual modelling lays the key foundation for the design and implementation phase of a system and also helps the users to intuitively express and validate their requirements and to prepare an explanatory document for the same.

Later, Kotsifakos and Ntoutsi (2005), pointed out the fact that patterns belong to multiple disciplines, like scientific, biological, mathematical, etc. and in order to manipulate these complex and diverse patterns, there is a need for a well-structured system which stores and represent these kinds of patterns properly. To achieve this aim, the author took few database models, like relational, object-relational and XML-based and compared them on the basis of a few criteria, like generality, extensibility and querying effectiveness.

Again, Tiwari and Thakur (2014b), designed a conceptual snowflake schema of pattern warehouse and derived patterns based on their structure out of a transactional dataset. The author used the extended snowflake schema for pattern warehouse representation and demonstrated it with a medical database.

Query Language Aspect

Terrovitis et al. (2006), established the relationship between the patterns and their corresponding raw data through a language, named Pattern Specification Language (PSL). Also, various operators like database operators, pattern base operators, cross-over database operators and cross-over pattern-base operators were introduced and these were combinations of traditional database operators.

Application Aspect

Tiwari and Thakur (2015), mapped the concept of pattern warehousing with the forensic databases. Their focus was to design a system which could store the patterns extracted out of the forensic databases in a non-volatile nature and perform analysis upon those patterns to develop a technique which perform forensic examination and analysis upon those patterns and generates the knowledge which directly helps in further investigative the decision-making process.

Gaps in Current Knowledge: Motivation towards Optimal Pattern Warehousing System

Discussion in the previous section indicates that there are still several flaws which hinder the efficient execution of a warehousing system for patterns. Some of these are:

- The various architectural components are missing from the architectures of different pattern management and warehousing systems, like there should be a repository which stores the schema of the pattern warehouse.

- Since, it is not necessary that patterns extracted from data warehouse after applying data-mining techniques are relevant, some patterns may be false frequently. So, there should be some optimizing technique that may filter out those irrelevant patterns.

- As the data sources get updated after a short interval of time which raises the need of updating the patterns within the pattern warehouse, it takes a long process to do so. In spite of that, some pattern predicting technique can be incorporated within the pattern optimization engine which can predict Future Frequent Patterns (FFP) based on the trends, thereby making the pattern warehouse optimal.

Proposed Optimal Pattern Warehousing System

Features of the System

The Optimal Pattern Warehousing System has several features which eliminate the shortcomings in the current pattern warehousing system and simplify the process of analyzing patterns by the business analyst. These features include:

- The OPWS architecture adds several important components which stabilizes the working of each tier and the architecture is expanded to six-tiers, making clear the whole process of extracting the patterns from raw data.

- The meta-pattern repository stores the schema of the pattern warehouse, enabling the pattern warehouse administrators to manage and modify the schema and syntactic structure of patterns effectively.

- The optimization engine in this system processes the patterns in order to mine the optimal ones and also predicts the patterns which encompass the feature of intelligence within this warehousing system.

Architectural Design

The architecture of the proposed Optimal Pattern Warehousing System is shown in Fig. 1 which is inspired from the three-tier data warehouse architecture as discussed by Han and Kamber (2006). The six main tiers of the proposed architecture are *Data Warehouse Server, OLAP Server for Pattern Generation, Adaptive Pattern Warehouse Server, OLOP Server for*

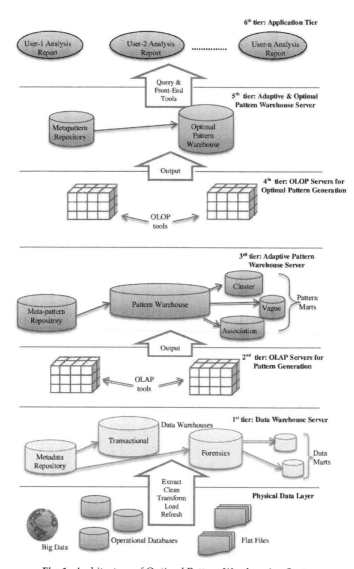

Fig. 1. Architecture of Optimal Pattern Warehousing System.

Optimal Pattern Generation, Adaptive and Optimal Pattern Warehouse Server and *Application Layer.*

1. *Physical Data Layer*: This is the lowermost tier that constitutes all the data files that will go on for processing on the higher levels of the system architecture. This layer collects all these data files from heterogeneous sources and preprocess them, i.e., cleans the data from noisy ones, extracts the relevant and non-redundant facts, transforms the data into a format that can be handled by the upper layers and finally loads the data into the data warehouses and refreshes it periodically.

2. *Data Warehouse Server*: The preprocessed data is transferred to this next higher tier that contains the repositories in which data is stored for multi-dimensional view in data warehouses.

3. *OLAP Server for Pattern Generation*: This tier holds the Online Analytical Processing tools that take the data of the data warehouse or data mart as input and applies analytical processing over it to mine patterns from this relational data. Various frequent pattern generation algorithms as described by Tiwari et al. (2010) are used by the OLAP engine, like Apriori, FP-growth, E-Clat, etc.

4. *Adaptive Pattern Warehouse Server*: In this tier, the patterns generated from the third-tier are stored in non-volatile form in a repository called Pattern Warehouse. Here also, each pattern warehouse is subdivided into pattern marts which contain the patterns of specific types. The meta-pattern repository comprises the data regarding schema of the pattern warehouse.

5. *OLOP Server for Optimal Pattern Generation*: This tier holds the Online Optimization Processing tools that takes the patterns of the pattern mart as input and apply optimization processing over it to extract the approximate best patterns from the pattern warehouse, thereby removing all false patterns and less accurate frequent patterns. Heuristics approaches, like GA, PSO, ACO, etc. are used for the optimization purpose.

6. *Adaptive and Optimal Pattern Warehouse Server*: This tier stores these Optimal Patterns (from previous tier) in a repository called Optimal Pattern Warehouse (OPWH). This storage space is updated on occurrence of new patterns.

Components of OPWS

The functioning of various layers has been discussed so far. Each tier contains few components which are responsible for the operational aspect of the Optimal Pattern Warehousing System. The following are the major components:

1. *Data Warehouse*: It is a repository which stores the data aggregated from discrete data sources in a multidimensional view. The size of the data repository spans from 100 GBs to TBs. The data stored in data warehouse is integrated, historical, time-variant and non-volatile. The data warehouse contains data of a specific domain, like shopping marts, hospitals, universities, etc.

2. *Data Mart*: It is a subset of the data warehouse to store the data concerning a specific department, for, e.g., data warehouse of an educational institution consists of data marts of every department, like account section, exam cell, training and placement, etc.

3. *Metadata Repository*: According to Ponniah (2001), this metadata repository contains schema-related information of the data warehouse, its different views, dimensions and derived data definitions, as well as data mart locations and its contents.

4. *Pattern Mart*: These are type-specific pattern databases which contain patterns of a specific class, i.e., association-rule patterns, clusters, decision-trees, etc.

5. *Meta-Pattern Repository*: This repository stores the schema of pattern warehouse and the syntactic structure of patterns belonging to different pattern classes.

6. *OLOP*: Online Optimization Processing is an optimization engine which applies heuristic algorithms over the patterns stored in the pattern marts and results in finding the optimal patterns which remain frequent for a long interval.

7. *Optimal Pattern Warehouse*: This is a repository which stores the optimal patterns. The end users finally dig into OPWH, apply queries to get analysis results for business decision making without knowing the background processing of the data.

User Scenario of the System

Implementation Aspect

In this scenario, the user interacts with the Optimal Pattern Warehousing System to extract the optimal patterns for efficient decision making. The objective is to provide a dataset and then perform mining and optimization operations over it and analyse the data for extraction of knowledge in the form of patterns. Figure 2 provides the flowchart of above scenario, which consists of following steps:

Step 1: Take a database as input (for, e.g., transactional, forensic, medical, etc.).

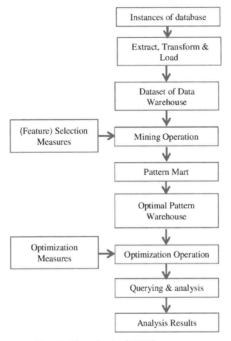

Fig. 2. Flowchart of IPWS process.

Step 2: Extract data, transform it into data warehouse format and load it into data warehouse.

Step 3: Now, select a dataset from data warehouse.

Step 4: Apply data mining operations over the dataset selected from data warehouse, like clustering, association rule mining (frequent pattern mining algorithm) and hesitation mining, etc. with the incorporation of appropriate mining algorithm.

Step 5: Store these generated patterns in pattern warehouse and each pattern of particular type will be stored in a specific pattern mart, i.e., association type, decision tress, etc.

Step 6: Take any one of the pattern marts and apply optimization algorithms upon it and generate high quality patterns.

Step 7: Store these optimal patterns in the optimal pattern warehouse.

Step 8: Apply queries and generate analysis reports.

Illustration

Now, illustrating the above scenario with the help of an example.

Step 1: A database of a *Store* is selected.

Step 2: Cleaned and transformed it into data warehouse format.

Step 3: Selection of dataset (Fig. 3).

Now, converting dataset into horizontal format for applying apriori algorithm as described in Han and Kamber (2006), Agrawal et al. (1993) to mine frequent patterns (Table 2).

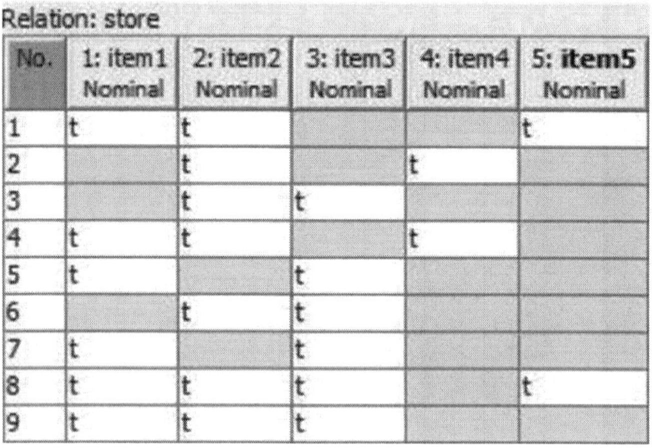

Relation: store

No.	1: item1 Nominal	2: item2 Nominal	3: item3 Nominal	4: item4 Nominal	5: item5 Nominal
1	t	t			t
2		t		t	
3		t	t		
4	t	t		t	
5	t		t		
6		t	t		
7	t		t		
8	t	t	t		t
9	t	t	t		

Fig. 3. Dataset of a store in binary format.

Table 2. Transactional dataset.

TID	Items
T1	I1, I2, I5
T2	I2, I4
T3	I2, I3
T4	I1, I2, I4
T5	I1, I3
T6	I2, I3
T7	I1, I3
T8	I1, I2, I3, I5
T9	I1, I2, I3

Step 4: Applying apriori as association rule mining algorithm as suggested by Pujari (2007) for finding frequent patterns with following inputs:

1. Dataset of 9 transactions.

2. Measure: Support value = 20 per cent of the no. of transactions.

Step 5: Following patterns are generated and are stored in the pattern warehouse. Since, all these patterns are of association type, so the association-type pattern mart contains patterns as shown in Table 3.

Table 3. Instances of association type pattern mart.

Size	Patterns	Support
1	I1	6
1	I2	7
1	I3	6
1	I4	2
1	I5	2
2	I1, I2	4
2	I1, I3	4
2	I1, I5	2
2	I2, I3	4
2	I2, I4	2
2	I2, I5	2
3	I1, I2, I3	2
3	I1, I2, I5	2

Step 6: Now, applying heuristic approach, i.e., genetic algorithm over pattern mart of step 5 with the following inputs:

1. Pattern mart
2. Fitness function will be calculated as:

Optimum Value = (0.5 * No. of items in the pattern) + (0.5 * Frequency_ count)

Where, Frequency_count is the number of times the pattern appears as a regenerated pattern after applying genetic algorithm over pattern set.

3. Threshold value: **Optimum value > = 1.5**

The optimum value is considered as 1.5 as this is a small dataset in which maximum frequent pattern size is 3. So the value is taken in accordance with the dataset and also single items will not constitute the category of patterns in order to eliminate those whose value is considered as 20 per cent of the maximum support calculated of the discovered pattern.

Before applying, the threshold value patterns generated are shown in Table 4.

Table 4. Instances of optimal pattern warehouse before applying threshold value.

PID	Pattern_kind	Size	Frequency	Pattern	Support	Opt_Value
P1	Old	1	2	I1	6	1.5
P2	Old	1	1	I2	7	1
P3	Old	2	1	I1, I2	4	1.5
P4	Old	2	1	I1, I5	2	1.5
P5	Old	3	2	I1, I2, I5	2	2.5
P6	New	3	3	I2, I3, I5	–	3
P7	New	2	1	I4, I5	–	1.5
P8	New	2	1	I3, I5	–	1.5
P9	New	3	1	I1, I3, I5	–	2

Step 7: Now after applying the threshold value, the high quality or optimal patterns are generated and stored in optimal pattern warehouse as shown in Table 5.

Table 5. Instance of OPWH.

PID	Pattern_Kind	Size	Frequency	Pattern	Support	Opt_Value
P1	Old	1	2	I1	6	1.5
P3	Old	2	1	I1, I2	4	1.5
P4	Old	2	1	I1, I5	2	1.5
P5	Old	3	2	I1, I2, I5	2	2.5
P6	New	3	3	I2, I3, I5	–	3
P7	New	2	1	I4, I5	–	1.5
P8	New	2	1	I3, I5	–	1.5
P9	New	3	1	I1, I3, I5	–	2

Step 8: Finally, application of the query over these patterns to generate either particular result or some analysis report or chart.

Proposed Genetic-based Algorithm for Finding Optimal Patterns (GBAOP)

Present section proposes a new genetic-based algorithm (as used in Step 6 of Section titled Illustration) for the generation of optimal patterns for Optimal Pattern Warehouse. This newly proposed algorithm utilizes the strength of genetic algorithm as discussed by Goldberg (2006), for ensuring the required quality and optimum level of the desired optimal patterns.

Workflow of Proposed Genetic-based Algorithm for Finding Optimal Patterns

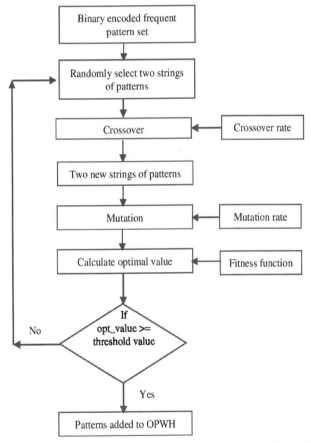

Fig. 4. Flowchart of genetic-based algorithm for finding optimal patterns (GBAOP).

Description of Flowchart

Step 1: Encode the frequent patterns mined by apriori algorithm in the dataset into binary format and give as an input to the algorithm.

Step 2: **Selection:** Select randomly any two strings for crossover. Selection of two mating strings of patterns can be performed by many ways. Here, random selection is used.

Step 3: **Crossover:** Here, single point crossover operator is used to genetically treat two pattern strings. In this step crossover rate is given as an input.

Crossover Rate: It is a parameter which adjusts the behavior of the genetic algorithm. Here, crossover rate is kept 95 per cent.

Step 4: After performing crossover of substrings, two new strings are generated which act as an input to mutation function.

Step 5: **Mutation:** It is a genetic operator which maintains the genetic diversity of strings from one population to next by flipping one or more bit values in the string from its initial state.

Mutation Rate: It is a measure of the resemblance that random elements of the string will be flipped into something else. Here, in this algorithm, mutation rate is kept very low at one per cent.

Step 6: Now, the optimal value of these two new strings are calculated which is a deciding factor for inclusion of these pattern strings into the Optimal Pattern Warehouse. Here fitness function will be given as input, which is used to calculate the optimal value of the pattern.

Fitness function: Optimum Value = (0.5 * No. of items in the pattern) + (0.5 * Frequency_count)

Step 7: Now, it is the final step in which if the optimal value is greater than a pre-defined threshold, the patterns will qualify to be stored in OPWH and the same process is repeated for all the pattern strings.

The Algorithm

The genetic-based pattern mining algorithm is divided into two phases—genetic-based pattern generation in which frequent patterns are generated from the old population of frequent patterns and second is prune-patterns, in which the optimal and future frequent patterns are extracted out of newly generated patterns. The algorithm is as follows:

Algorithm: Genetic-based algorithm for finding optimal patterns

Inputs: Binary encoded frequent pattern set, crossover rate and mutation rate.

Output: Optimal and future frequent patterns for Optimal Pattern Warehouse.

Phase I: Genetic-based pattern generation (Pattern-set, No. of Patterns, No. of Items)

Initialize: i:=1, k:=1 n:= no. of patterns, j:= total no. of items

P_{n*j}:= contains all patterns in bit representation

NP_{n*j}:= contains all high-quality patterns

Fv_{n*1}:= 0, frequency of occurrence of pattern

Pt_{n*1}:= 0, type of pattern, i.e., 0 denotes old pattern, 1 denotes new pattern

for all patterns $P_{n*j} \in P$

while $(i <= upper\_bound(n/2))do$

begin

     $Parent1 := P_{i*j}$;   // *Selection of two patterns*

     $Parent2 := P_{(n-(i-1))*j}$;

      if parent1 equals parent 2

        $NP_{:=} \cup NP_{k*j} = Parent1$ or $Parent2$;

        call $Update\_pattern\_vector(NP_{k*j})$;

        $k := k+1$;

     else   //*Crossover*

        $N' P_{k*j}, N' P_{(k+1)*j} := crossover(Parent1, Parent2,$

$lower\_bound(j/2))$;

        $NP_{:=} \cup N P_{k*j} := mutate(N' P_{k*j}, 1)$;   // *Mutation*

        $NP_{:=} \cup N P_{(k+1)*j} := mutate(N' P_{(k+1)*j}, 1)$;   // *Mutation*

        call $Update\_pattern\_vector(NP_{k*j})$;

        call $Update\_pattern\_vector(NP_{k+1*j})$;

        $k := k+2$;

end

Update_pattern _vector(NP$_{m*n}$)

for each new $NP_{m*n} \in NP$

  if $NP_{m*n} \in P$ or $NP_{m*n} \in NP$

   $Fv_{m*1} := Fv_{m*1} +1$;   // *Update frequency of occurrence of pattern*

  else

   $Fv_{m*1} := Fv_{m*1} +1$;   // *Update frequency of occurrence of pattern*

$Pt_{m*1}:=1$;   // *Update of new pattern*

return

Phase II: Prune_Patterns (NP, k, j, Fv)

Initialize: i:= 1.

for all $np_{i*j} \in NP$

while (i< k) do

begin

$x:= count\_no\ of\ 1's\ in(\ np_{i*j})$;   // *Calculating no. of items in a pattern*

$op\_value:= (\ 0.5 * x) + (\ 0.5 * Fv_{i*1})$; // *Calculating fitness value of a pattern*

if op_value < 1.5

discard np_{i*j}

end

Working of Algorithm GBAOP

The flow of the algorithm works as explained in Fig. 4. Here, the author demonstrates the step-by-step working of the algorithm.

Inputs: The dataset given in Table 2 is considered over which apriori algorithm is applied to get frequent patterns (Table 3), which becomes the input for this algorithm and now it is encoded in binary format (see Table 6).

Table 6. Binary encoded frequent pattern-set.

Pattern No.	Item1	Item2	Item3	Item4	Item5
T1	1	0	0	0	0
T2	0	1	0	0	0
T3	0	0	1	0	0
T4	0	0	0	1	0
T5	0	0	0	0	1
T6	1	1	0	0	0
T7	1	0	1	0	0
T8	1	0	0	0	1
T9	0	1	1	0	0
T10	0	1	0	1	0
T11	0	1	0	0	1
T12	1	1	1	0	0
T13	1	1	0	0	1

Phase I: Genetic based pattern generation (Pattern-set, No. of Patterns, No. of Items)

Initialize: i:=1, k:=1, n:=13 (no. of transactions), j:=5 (no. of items)

P_{13*5}:= contains pattern-set (Table 6).

NP_{13*5}:= null, Fv_{13*1}:= 0, Pt_{13*1}:= 0

while i <= 7

Parent 1:= T1

Parent 2:= T13

Since,   Parent 1 ≠ Parent 2,

Therefore,   Crossover (Parent 1, Parent 2, 2)

Now,

T1	1	0	0	0	0
T13	1	1	0	0	1

NP1	1	0	0	0	1
NP2	1	1	0	0	0

(bit to be mutated)

In next step,   Mutation is performed on both string, as mutation rate given is one per cent; therefore, only 1 bit gets flipped. Suppose last bit is mutated.

So,

NP1	1	0	0	0	0
NP2	1	1	0	0	1

Now,   update_pattern_vector(NP1)

Since,   NP1 ∈ P   i.e., old population

Therefore, Fv_{1*1} := 0+1 = 1

Again,   update_pattern_vector(NP2)

Since,   NP2 ∈ P   i.e., old population

Therefore, Fv_{2*1} := 0+1 = 1

Now, in the same way other strings will get selected and populate the new pattern-set (NP) as shown in Table 7:

Table 7. New pattern-set.

PID	Pattern_Type	Frequency	Pattern				
NP1	0	2	1	0	0	0	0
NP2	0	1	0	1	0	0	0
NP3	0	1	1	1	0	0	0
NP4	0	1	1	0	0	0	1
NP5	0	2	1	1	0	0	1
NP6	1	3	0	1	1	0	1
NP7	1	1	0	0	0	1	1
NP8	1	1	0	0	1	0	1
NP9	1	1	1	0	1	0	1

Now, Phase II of the algorithm is executed in which optimum value is calculated to filter out the optimal patterns.

Phase II: Prune_patterns (NP, k=10, j=5, Fv)

Here,	Optimal_value of each pattern is calculated
Initialize,	i:=1,
while	i < 10
	x: = no. of 1's in (NP1)
	x: = 1
	opt_value := (0.5 * 1) + (0.5 * 2) =1.5
Since,	opt_value > threshold, i.e., 1.5
Therefore,	NP1 is included in IPWH.
Again,	i:=2,
while	i < 10
	x: = no. of 1's in (NP2)
	x:=1
	opt_value := (0.5 * 1) + (0.5 * 1) =1
Since,	opt_value < threshold, i.e., 1.5
Therefore,	discard NP2.

Hence, in the same way optimal values of the other patterns is calculated and optimal patterns are therefore stored in Optimal Pattern Warehouse as shown in Table 5.

Analysis of Generated Patterns and Execution Time of Algorithm

The analysis of the high quality and optimal patterns stored in optimal pattern warehouse can is explained by the charts in Fig. 5 and Fig. 6.

In Fig. 5, horizontal axis represents size of patterns and vertical axis signifies possible support values of patterns. The frequent old and new patterns are plotted in the above chart based upon their size and respective support which shows that while considering a dataset of five items and nine transactions, patterns of size two are having decent support value, are frequent and listed in future frequent patterns.

In Fig. 6, the horizontal axis denotes the size of patterns and vertical axis represents the frequency of occurrence of frequent patterns. The frequent old and new patterns are plotted in the above chart based upon their size and respective frequency_count which indicates that a number of patterns of size two and three are frequent, i.e., their frequency of occurrence within the transactions is high, which means that items in these patterns are more likely to be purchased by customer. Now, Table 8 describes different pattern-sets to in order to analyze the running time taken by the algorithm to extract optimal and future frequent patterns out of them.

Here, in Fig. 7 the graph between different pattern-sets and execution time of algorithm is drawn. The pattern-sets contain varying numbers of frequent patterns over which algorithm is executed to extract optimal patterns. The execution time of algorithm is the time of extracting the optimal patterns from these pattern-sets. The analysis of the graph shows that the execution time of the algorithm is growling linearly when implemented over pattern-sets containing varying number of patterns.

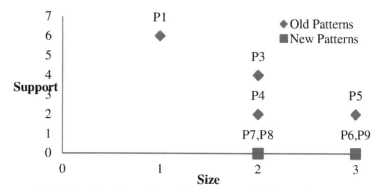

Fig. 5. Chart showing patterns with respect to their size and support.

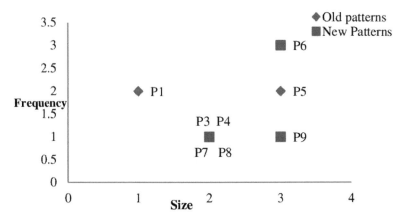

Fig. 6. Chart showing patterns with respect to their size and frequency of occurrence of patterns.

Table 8. Different frequent pattern-sets.

Pattern-set	No. of Patterns
PS1	20
PS2	50
PS3	100
PS4	1000
PS5	10000

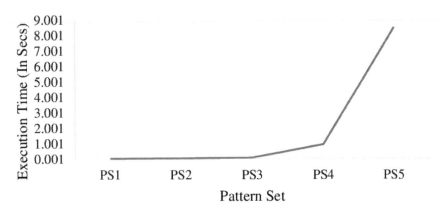

Fig. 7. Graph showing the execution time of algorithm for different pattern-sets.

Critical Analysis of Existing Pattern Warehouse Architecture and Proposed Architecture

After discussing the architectural and conceptual aspects of proposed optimal pattern warehousing system along with the illustration that describes operational scenario of system, the author compares the features of various existing pattern warehousing system architecture with proposed architecture. Table 9 shows a comparative analysis of the existing pattern warehouse architectures with the proposed optimal pattern warehouse architecture.

Table 9. Existing pattern warehousing system vs. optimal pattern warehousing system.

Features	Existing Pattern Warehousing System	Optimal Pattern Warehousing System
Processing Engines	OLAP (Online Analytical Processing)	Both OLAP and OLOP (Online Optimization Processing)
No. of Tiers	3 or 4 tiers	6-tiers
Outcome	Pattern Analysis	Optimal Pattern Analysis
Meta-pattern Repository	Absent	Present
Concept of Pattern Mart	Unspecified	Specified (Pattern Mart stores the pattern of specific pattern-type)
Reliability of Patterns	Presence of false frequent patterns	More reliable, potential and future frequent patterns (FFP)
Repository for Storing Optimal Patterns	Not present	Optimal Pattern Warehouse
Techniques for generating patterns	Traditional approaches for finding frequent patterns	Genetic-based approach for finding optimal patterns

Key Findings of the Research Work

This section gives the details regarding major findings and outcomes of the present contribution by research.

1. Various new and necessary components were missing from the previous architectures which are now incorporated in the proposed Optimal Pattern Warehouse Architecture.

2. The proposed genetic-based Optimal Pattern Warehousing System provides a repository where optimal patterns are stored in a persistent manner.

3. Among various heuristic approaches available in the literature, the author discovered that genetic algorithm is less complex to couple

with complicated pattern-mining algorithms and also here, the convergence condition is user-specified which is a number of frequent patterns, i.e., until all the frequent patterns are selected and genetically treated by the algorithm and it can be represented as:

$$\text{No. of iterations} = |n|/2$$

4. As genetic algorithms deal with the trends and patterns, i.e., two strings of old population are genetically treated to get strings for new population and these two new offsprings contain features of the previous population. In the same way, in the proposed approach, there is an old population of frequently generated patterns out of which most optimal patterns and some new future frequent patterns (FFP) are now generated by applying genetic algorithm.

5. Moreover, the algorithm is executing in polynomial time with different sizes of pattern-sets. Therefore, the proposed genetic-based algorithm for optimal patterns is having a time complexity of $O(n)$.

Conclusion and Future Work

This chapter presents the framework for genetic-based Optimal Pattern Warehousing System. The efficiency of the proposed Optimal Pattern Warehousing System depends on efficient generation, storage and management of optimal patterns in the optimal pattern warehouse. This warehousing system supports the whole lifecycle of optimal patterns from their generation to retrieval. The system is optimal in the sense that it predicts the patterns on the basis of trends that are frequent when database updates. Besides, the author also addresses the issues related to updating the pattern warehouse. Future research includes implementation-specific issues regarding defining the generalized schema of optimal pattern warehouse, specific query language for optimal pattern extraction, probability of imposing semantic constrains on patterns and other operations related to patterns.

References

Agarwal, V., A. Tiwari, R. K. Gupta and U. P. Singh. (2018). Discovering optimal patterns for forensic pattern warehouse. pp. 101–108 *In*: Choudhary, R., J. Mandal and D. Bhattacharyya (eds.). Advanced Computing and Communication Technologies. Advances in Intelligent Systems and Computing. Vol. 562, Springer Nature, Singapore.

Agrawal, R., T. Imielinski and A. Swami. (1993). Mining association rules between sets of items in large databases. pp. 207–216. *In*: Proceedings of the 1993 ACM SIGMOD International Conference on Management of Data, ACM, New York, USA.

Bartolini, I., E. Bertino, B. Catania, P. Ciaccia, M. Golfarelli, M. Patella and S. Rizzi. (2003). Patterns for next-generation database systems: Preliminary results of the PANDA project. *In*: Proceedings of the 11th Italian Symposium on Advanced Database Systems, SEBD, Cetraro (CS), Italy.

Catania, B., A. Maddalena and M. Mazza. (2005). PSYCHO: A prototype system for pattern management. *In*: Proceedings of the 31st VLDB Conference, Trondheim, Norway.

Goldberg, D. E. (2006). Genetic Algorithms in Search, Optimization and Machine Learning, Pearson.

Han, J. and M. Kamber. (2006). Data Mining: Concepts and Techniques. Second Edition, Morgan Kaufmann Publishers, San Francisco, Elsevier.

Kotsifakos, E. and I. Ntoutsi. (2005). Database support for data mining patterns. *In*: Proceedings of the 10th Panhellenic Conference on Advances in Informatics.

Kotsifakos, E. E., I. Ntoutsi, Y. Vrahoritis and Y. Theodoridis. (2008). Pattern-miner: Integrated management and mining over data mining models KDD. Proceedings of the 14th ACM SIGKDD International Conference on Knowledge Discovery and Data Mining, pp. 1081–1084.

Ponniah, P. (2001). Data Warehousing Fundamentals: A Comprehensive Guide for IT Professionals, Wiley-India.

Pujari, A. K. (2007). Data Mining Techniques. Third Edition, Universities Press, India.

Rizzi, S. (2004). UML-based conceptual modelling of pattern-bases. *In*: Proceedings of the International Workshop on Pattern Representation and Management (PaRMa), Heraklion, Hellas.

Terrovitisa, M., P. Vassiliadis, S. Skiadopoulos, E. Bertino, B. Catania, A. Maddalena and S. Rizzi. (2006). Modeling and language support for the management of pattern-bases. Data & Knowledge Engineering, Elsevier pp. 368–397.

Tiwari, A., R. K. Gupta and D. P. Agrawal. (2010). A survey on frequent pattern mining: Current status and challenging issues. Information Technology Journal 9: 1278–1293.

Tiwari, V. and R. S. Thakur. (2014a). P2MS: A phase-wise pattern management system for pattern warehouse. International Journal of Data Mining, Modeling and Management, Inderscience 5: 1–10.

Tiwari, V. and R. S. Thakur. (2014b). Contextual snowflake modeling for pattern warehouse logical design. Sadhana Academy Proceedings in Engineering Science, Springer 39: 15–33.

Tiwari, V. and R. S. Thakur. (2015). Improving knowledge availability of forensic intelligence through forensic pattern warehouse (FPW). pp. 1326–1335. *In*: Khosrow-Pour, M. (ed.). Encyclopedia of Information Science and Technology, IGI Global. Hershey.

CHAPTER **9**

Controlling Social Information Cascade
A Survey

Ragia A. Ibrahim,[1,*] *Aboul Ella Hassanien*[2] and
Hesham A. Hefny[1]

Introduction

Recently, there has been a huge information flow through social networks. Individuals communicate and receive information which may lead to a change in the individual state. The state could be an opinion or an action, such as adopting a new product or following a specific political party. People interact and socialize via social networks by attending or giving importance to a special event, a web page or maybe a group. In addition, marketing is not an independent action anymore. It can be viewed as a social network affecting the customer's decision to buy or not. Therefore, analyzing and controlling the influence over social networks is important.

The spread and elimination of information propagation over social networks has become a critical issue from several points of view. Influence maximization started mainly to facilitate viral marketing. The early studies focused on person-to-person influence. Domingos and Matt Richardson

---
[1] Institute of Statistical Studies and Research (ISSR), Cairo University, Egypt.
[2] Faculty of Computers and Information, Cairo University, Egypt.
[3] Scientific Research Group in Egypt, Cairo university, Egypt.
Emails: h.hefny@ieee.org; aboit@gmail.com
* Corresponding author: ragia.ibrahim@gmail.com

(2001) thought of viral marketing or by word of mouth as an epidemic spread. Marketing is not an independent action anymore; it can be viewed as a social network; for example, such as in hybrid seed corn in Iowa that was studied by (Ryan and Gross 1943), or drug adoption (Coleman et al. 1970, Katz et al. 1966).

Maximizing information cascade becomes inevitable in many cases, such as in a emergency situation, where information spreading could save a person's life. During 2011, the East Japan earthquake and tsunami, where Twitter and Facebook served as a lifeline for directly affected individuals (Peary et al. 2012). In addition, people's behavior before and after the event can be inferred. Social media was used effectively during and after this earthquake, and the situation was revealed on a blog. How people react on and after an earthquake is viewed on the blog in case of any disaster (Sakaki et al. 2011). In an unexpected critical situation, such as when Iraq army capitulated to ISI militants in four cities, the social networks started to tweet and post about canceling airlines trips and changing plans regarding the new situation (The Guardian March 2011).

Also, influence can affect a person's social decisions, One of the most well-known examples of influence propagation is Ted Williams, called 'Golden Voice'. He turned from a homeless man in Columbus, Ohio into a saved person. A reporter recorded a video for the homeless Ted. The video was re-posted to YouTube where it received 11 million hits in months and Williams received several job offers (Foxnews March 2014).

Many applications, actions and events have been built based on the implicit notation of social influence between people. According to (Zafarani et al. 2014) social networks can be classified as an explicit network where individuals and their relations are declared on implicit network. In the explicit network, diffusion depends on a global decision as in herding behavior or local decision (depends on the neighbors choices), as in information cascade. Implicit network means that the graph is not explicitly constructed, but remains implicit. Incrementally construct nodes in the implicit graph as we search for a solution in the hope of finding a solution without ever generating all the nodes. Diffusion of innovation and epidemics is classified as propagation on implicit network.

Collective action integrates the network effect on both the network global level and the node local level, for example, when organizing a protest. If a huge number of people participate, everyone in the society will benefit, but on the other hand, if a few hundreds only show up, then the demonstration will arrest them. Thus, in collective behavior the benefits occur only when enough number of users adapt specific opinions or actions. Individuals in such a model will adopt the idea if at least k users of their neighbours adopt it (Zafarani et al. 2014).

Rumors cascades elimination is needed as daily rumors are being spread around. This kind of information causes frustration, such as in fake news like the death or marriage of celebrities, but also, it could cause panic and unexpected individual behavior, such as in spreading Ebola panic via Facebook (THEEVERGE October 2014) or it might affect the economic currency planes, such as the fake news about withdrawal of some Indian banknotes from circulation (BBC December 2016). In other cases, it is aimed to digitally alter a fake info, such has in a statement claiming that Abu Bakr al-Baghdadi had died (The Independent June 2011).

This survey represents recent work in controlling information spread—in both cases, maximizing and maximization. This chapter is organized as follows: First, presents cascading models and recent work in influence maximization. Second, discusses the recent work on influence minimization and finally, presents applications and challenges presented.

Influence Maximization Problem

Several studies showed influence of power in decision making. David Krackhardt created an advice network by collecting data on manufacturing organization. The network number of nodes was approximately 100, representing the answer to a question, "To whom do you go for help or advice?". Also, the network includes the person level in the company hierarchy, age and degree. Krackhardt found that some persons on level 2 had more influence than persons in the top hierarchical level (Krackhardt 1987).

A landmark study by Coleman, Katz and Menzel (Coleman et al. 1966) documents that for social contagion (diffusion of innovation) doctors decided to adopt a new drug influenced by people they were connected to, considering social and professional connections. They found that the number of doctors who had not been named as a friend or as an advisor (disconnected) was much lower in the beginning but gradually the rate of adoption increases and the gap between the rates decreases.

Ryan and Gross examined the diffusion of innovation in 1943 (Ryan and Gross 1943) when the farmers adopted a new hybrid seed. Ryan and Gross were differentiated between adopters by classifying them into Innovators, early adopters, early majority, later majority and laggards who didn't adopt the new technology at all (hyper red corn). Later on, Griliches (Griliches 1957) expanded the notion of hybrid spread diffusion by examining the adoption patterns in other agricultural regions of the United States and showed that it took an 'S' shaped plot. Several approaches have been reported for this influence-maximization problem.

First Influence Maximization Model

The first proposed model was by Domingos and Matt Richardson (2002) and Domingos and Matt Richardson (2001). Matthew Richardson and Pedro Domingos defined IM model in terms of Markov Random fields' formulation. Each node 'X' in Richardson and Domingos' model has a value 'I' that indicates if the node has bought the product or not, each is independent of its neighbor's actions and marketing action 'M' consists of $(M_1, M_2, ..M_n)$. Richardson and Domingos define the IM problem as finding the set of 'M' that maximize the revenue obtained from the result of 'X'. The model is very simple and it does not include product attributes. In addition, computation of revenue includes marketing cost, discount and Bayes model which were used to simplify the computation.

David Kempe, Jon Kleinberg and Eva Tardos defined discrete diffusion models over social networks G = (V, E) where 'V' is social network nodes and 'E' is the set of edges. The diffusion process as defined on Kempe's model depends on the initial seed set of nodes 'S' that started the diffusion. Nodes are activated directly via stochastic models from the seed set and the influence spread function $\sigma(S)$ is defined as the expected number of diffusion-activated nodes by seed set 'S' at each end of the diffusion process (Kempe et al. 2003). The major defined influence-stochastic models by (Kempe et al. 2003) are Linear Threshold model (LT) and Independent Cascade model (IC).

Independent Cascade Model (IC)

Independent cascade IC model is a stochastic (prediction) model that depends on graph structure, whether it is highly connected or a sparse graph. Cascading means the behaviour of being active or inactive that results from the inference (function) based on neighbours or earlier individuals. Thus cascading is based on the influence of neighbours that affects each node or individual to be earned in one of two states—active or inactive (Guille et al. 2013).

Firstly, the seed nodes are selected are in an active node (infectious). These seeds trigger information spread or the virus over the network. Then, each node or individual has the property to infect its neighbors according to their tie (edge) to the node. Thus IC uses diffusion probability associated with each edge and at each iteration, each active node uses that probability once to activate its neighbors. Accordingly, for nodes (u,v) if 'u' is active and 'v' is not, the activation succeeds depending on node 'v' and 'u' and other nodes that already attempted to change 'v' but failed. Once 'v' is activated, it can activate its neighbours too (Kleinberg 2007, Guille et al. 2013).

Linear Threshold Model (LT)

This is a stochastic diffusion model which depends on the graph structure as in IC. It also depends on the degree of influence for each edge and threshold for each node. LT assumes first that the influence weight denoted by $b_{v,w}$ is not negative on each directed edge (v, w). In addition, the sum of all weights to a specific node 'w' belongs to 'v' set of nodes is less or equal to 1. Second, each node 'v' is uniformly random $(0, 1)$ to choose function θ_v that specify fraction of 'v' neighbors should adopt the new behaviour before 'v' does (Kleinberg 2007).

In LT, the activation proceeds continually. At LT, firstly each node is inactive. Then, the seed nodes select and switch from an inactive state to active one. Continuously at each iteration '(t)' the inactive node changes its state to active state if $\sum_{(\text{active } w \in N (V)} b_{(v, w)} \geq \theta_v$. This means that the sum of influence coming from its neighbors is greater than its own influence threshold. Small θ indicates that the node tends to adopt a new behavior. The process ends when no more nodes get infected. Recently researchers have investigated that a node might adopt the behavior if a few neighbors adopt it but might resist it if many neighbors do so (Kleinberg 2007, Easley and Kleinberg 2010, Guille et al. 2013).

Influence Maximization Hardness

Given a social network G, a diffusion model with certain parameters and budget k maximization function $\sigma(S)$, as defined by Kempe et al. (2003) finds seed set S that has a number of k nodes such that the influence spread S is maximized. Kempe et al. (2003) proved that $\sigma(S)$ is monotone, non-negative and submodular function. Therefore, the theorem of Nemhauser, Wolsey and Fisher (Nemhauser et al. 1978) was applied. If a function is non-negative, monotone and submodular, the greedy algorithm provides a $(1-1/e) \cong 63$ per cent approximation of the optimal value. Authors deduced that the maximization problem could be solved via greedy algorithm. The greedy algorithm is a solution within 63 per cent of linear time.

Damon Centola and Michael Macy (2007) stated that IC is a simple contagion while LT is a complex contagion that needs social confirmation from multiple sources. In the independent cascade model, they simulated information diffusion on artificial generated network that consisted of two types of ties—strong ties within groups and weak ties between the groups. They found that the weak ties were necessary for information spread rate, but long ties do not always facilitate the spread of complex contagions and can even prevent the diffusion.

Algorithm 1. General Greedy Algorithm

Input: Graph (G) and k: number of initial nodes

Output: selected seed set S

initialization $S = \phi$

FOR i=1 to k

  FOR $v \in V$

    FOR $j := 1$ to \$R\$}

      v_{mg} += F(S)

    ENDFOR

    $v_{mg} = v_{mg} \backslash R$

  ENDFOR

        STATE $S \rightarrow S \backslash \cup$ argmax$_v$

ENDFOR

  return(S)

General Threshold Model

Kempe et al. (2003) defined the General Threshold model as follows—each node v $\in V$ has a threshold function $F(v)$: $2^v \rightarrow (0,1)$ wherein each node v selects a threshold $\theta_v \in (0,1)$ uniformly at random. If the set of active nodes at t-1 is S and $F(S) \geq \theta_v$ then at step t node, 'v' is activated. The reward function r is defined as $r(A(S))$. If the set $A(S)$ is the final set of activated node, then the reward of the seed set S is $r(A(S))$ and the generalized influence spread is $\sigma(S) = E[r(A(S))]$.

Elchanan Mossel and Sebastien Roch (2007) proved and extended Kempe et al. (2003) General Threshold Model, Mossel and Roch Theron stated that in the General Threshold node for every $v \in V$, $f_{V(.)}$ is monotone and submodular for $F_v(\theta) = 0$, and the reward function r is monotone and submodular than the general influence spread function $\sigma(S)$.

As Kempe et al. (2003) proved IM is and NP-hard and can be approximated by using the greedy algorithm and Monte Carlo simulation estimates $\sigma(S)$ with high probability of arbitrary accuracy. But greedy algorithm suffers from several inefficiencies—it takes days to find seed set S almost 50 elements for a network of 30 k nodes using a highly facilitated machine. Too many evaluations for influence spread given the graph of number of 'n' nodes means that for each round, each node is evaluated. Totally '$O(nk)$' evaluations and each evaluation is #P-hard for both IC and LT models (Chen et al. 2010).

Scalable Influence Maximization

Regarding the greedy inefficient a scalable influence maximization, the scalable algorithm aims to reduce the number of influence spread evaluation and uses batch computation of influence spread. The following is a brief description of scalable presented algorithms.

Cost-effective Lazy Forward Optimization CELF

Leskovec et al. showed that greedy is slow since it evaluated marginal gain in each iteration, replacing the sensor and recalculating everything and this happened every time the sensor is added. In addition, greedy always prefers a more expensive sensor with reward 'R' to a cheaper sensor with reward 'R'. Thus Leskovec et al. (2007) developed an algorithm. CELF (cost-effective lazy forward-selection) that performed near optimal solution 700 times faster than greedy algorithm. CELF solution is getting the maximum of one of the two calculated rewards. Reward R(A): benefits for the cost and 'R(B)' benefit without calculating the cost. The algorithm works as follows—calculate rewards and took the largest one. The remaining sensors are then reevaluated and perform lazy evaluation by taking the next ordered sensor in the priority queue without recalculating the reward from them. Because of submodularity guarantees, the marginal

---

Algorithm 2. Lazy Evaluation with General Greedy Algorithm

Input: Graph (G) and k: number of initial nodes
output: {selected seed set S of length K
initialization $S = \phi$, $PQ = \phi$, iteration → 1
  FOR $v \in V$ do
    $v.mg → F(v, S)$
    i → 1
  ENDFOR
  WHILE (iteration <= K)
      pop max element v of Q
        IF $v.i$ = iteration
          $S → S \cup \{v\}$
        ELSE
            $v.mg → F(U,S); v.i →$ iteration
            reinsert v into PQ}
        ENDIF
  ENDWHILE
    return(S)

---

benefits decrease with the solution size. Thus, it will give dimensionality return more than the next one to it.

This algorithm could be implemented via Priority Queue 'PQ'. Each node has got two proprieties—firstly, the marginal gain 'mg' of the node and secondly, the iteration 'i' number. The function $F(v, S)$ calculates the marginal gain for the node 'v' given 'S'. The first time 'S' will be empty and all nodes the marginal gain will be calculated. By taking the top of the queue and examining its marginal gain 'mg', if it has the marginal gain of the current round, then it is the new seed. Often the top elements in round 'k-1' remain at top in round 'k'.

Optimized Cost Effective Lazy Forward Optimization CELF++

Goyal et al. (2011) made further optimization for CELF. For each node 'u' Goyal et al. defined two marginal gains: The first marginal gain mg $u.mg\_1$ = $MG(u \backslash S)$, where 'S' is the current seed set. Also defined node is the previous best marginal gain u_prevbest is node max best marginal gain in current iteration among nodes have been seen before 'u'. The second node 'u' marginal gain is $u.mg_2$ = $MG(u \setminus S) \cup$ u_prevbest. Goyal et al. defined node flag as $u.flag$ that reflects the iteration where node 'u' is last updated. In CELF++, if node 'u' previous best is current iteration, there is no need to calculate $u.mg_2$ in the next iteration. In addition $m.mg_1$ and $u.mg_2$ are calculated in the same run in Monte Carlo simulation. The optimized algorithm achieves up to 50 per cent additional improvement in terms of efficiency.

Batch Computation of Influence Maximization

Cohen (1997) studied the size estimation framework in a directed graph estimate reachability for each node using transitive closure. Cohen generated a random life edge graph. Then he used batch computation to estimate the number of all reachable nodes from every node before performing distinct counting to ensure that descendants reached by multiple paths are not double counted. In conflict with CELF Chen et al. (2010) proposed mixed greedy, where batch-based computation and user's reachability in the first run in the other runs uses CELF. Chen et al. proposed the scalable method that's faster than the existing greedy algorithm for Maximum Influence Arborescence (MIA) model. Chen et al. invented a heuristic scalable MIA algorithm for 'IC' on the directed graph. Firstly the algorithm evaluates the influence of each node when the seed set is empty. Then the algorithm iterated two steps till finding the 'k' seeds. Chen et al. focused on local regions and used the batch marginal gain update, where node 'u' represents the new seed set in set 'S'. The marginal gain update is based on linear relation (the graph converted into a tree) between candidate seed 'u' and root 'v', to find the local influence region for node 'v'.

- Find the maximum influence path 'MIP' from node 'u' (for all nodes in the graph) to 'v', given the largest marginal influence. For every node, find maximum influence 'MIP', which is computed via a Dijkstra shortest path algorithm. The 'MIP' represents maximum propagation property '$pp(P)$' of the path 'p'. MIP's that is less than specific θ are ignored. The added node 'u' to the seed 'S' is an element of maximum influence in arborescence MIIA. MIIA estimates the influence to 'v' from other nodes.

- Update the MIA after selecting 'u' as a seed, where the union is of MIP. Authors found updating process is a feature in their system since updating was done locally. The update is for all 'w' nodes to 'v'.

Chen et al. were close to greedy algorithm results and much faster at that.

The Simpath Algorithm

Amit Goyal, Wei Lu and Lakshmanan introduced Simpath (Goyal et al. 2011) for Linear Threshold LT model. The SimPath algorithm optimized the previous algorithms by enumerating simple paths started from a node 'v'. Simpath, the algorithm, is based on CELF under the LT model, incorporating two optimizations to reduces the number of spread estimation calls. Optimizations made by Simpath are called Vertex Cover Optimization and Look Ahead Optimization. The Vertex Cover Optimization is used to improve the efficiency of the first iteration where the spread of the node can be computed by enumerating the simple paths started from node 'v' to its neighbors at the induced sub-graph of '$V–S + u$'. The total influence is estimated by the simple paths from seed set 'S'. The Look Ahead Optimization improves the efficiency of the subsequent iterations since as the seed set 'S' grows, the time spend to estimate enumerated paths increases. The algorithm avoids this reputation of using a look ahead parameter 'l'. However, the greedy algorithm's quality of influence spread is guaranteed to be within 63 per cent of the optimal, but these scalable models' is not (Li et al. 2015).

Community-based Influence Maximization

Yu Wang et al. (2010) proposed a new algorithm named community-based greedy algorithm CGA that finds top 'K' influential nodes in a large MSN data set. The algorithm was applied to Independent Cascade model and they extended IC to include the edge weight. The CGA portion of the graph was extended into communities since that will reduce the search space of the problem of finding influential users. Yu Wang et al. assumed that the influence of a user was limited to each community. The community-based greedy algorithm carried out Monte Carlo simulation within each community, but not for the whole network. In addition, CGA selected each

---

Algorithm 3. LIM Algorithm

Input: Graph (G)

Output: selected seed set S

initialization $S = \phi$;

Partition G into i sub graphs

RImportance (G_i)

FOR i = 1 to k

  FOR $v \in G_i$}

    $v_{mg} = F(v_{ci})$}

      \STATE {$U_{ci} = $ argmax(C_i)}

  ENDFOR

$S \rightarrow \cup F(U_{ci})$}

ENDFOR

  return(S)

---

influential node based on a unified optimization formula, which required the information for all communities stored in a unified space.

Ragia et al. (Ibrahim et al. 2016) proposed Local Information Maximization (LIM), which uses the Walktrap well-known algorithm to separate the graph into sub-graphs regarding communities. Then the select 'k' initial set was distributed among the communities depending on the relative importance that uses community density and connectivity.

This is formed if and only if the new seed set can generate better influence spread than the old one conducting the IC model.

Simulated Annealing-based Influence Maximization

Jiang et al. (2011) reported significant speedups by using simulated annealing SA which does not go through all the nodes to find the next seed as in greedy, but initiated the seeds set by randomly selecting 'k' nodes guided with a heuristic. A new seed set is formed if and only if the new seed set can generate better influence spread than the old one conducting the IC model.

Influence Minimization Problem

Collective action integrates the network effect on both network global level and node local level; for example when organizing a protest, if a huge number of people participate, everyone in the society will benefit, but on the other hand, if a few hundreds only show up, then the demonstrators will be arrested. Thus, in collective behavior, the benefits occur only when enough number of users adopt specific opinion or action. Individuals in such a model will adopt the idea if at least '*k*' users and their neighbors adopt it (Zafarani et al. 2014).

Influence of viral marketing in the early studies focused on person-to-person influence, such as was seen in the hybrid seed corn in Iowa that was studied by Ryan and Gross (1943) and in drug adoption by Coleman et al. (1966) and Hagerstrand (1970). Recently in February 2013, the Oreoaa Sandwich Cookie Company tweeted, during a power failure, in the Super Bowl XLVII which is an example of maximum benefit gained at minimum cost (Business Insider March 2013).

Considering the misleading information (rumor) spreading over the social network, the target is stopping this cascade. An example of rumor information was that Facebook was going to charge (CNBC September 2015) Justin Bieber who was a famous victim of death rumour. His death was reported roughly every two weeks, according to Synthesio (December 2013) which is a global social media and analysis company. In addition, the rumor of the death of many famous actors or political figures, for example, the death of the formal President of Egypt, Hosni Mubarak, was spread over the social networks many times (CNN January 2012).

Influence Minimization via Edge Removal

Several studies have targeted influence propagation minimization via several methods, e.g., edge deletion, node deletion, computed cascade and blocking nodes.

Eliminating cascade among networks could be a priority, especially during troubled times, such as an epidemic spread over water drinking network, or false rumors or bad information cascade.

The main advantage of edge removal versus node removal is in removing the node and all its linked edges from the network. On the other hand, removing set of edges, such as the inflow-specific person on Twitter, resulted in removing only the tie between two actors. Eliminating cascade via edge blocking has been studied for IC model to minimize the spread of unwanted information (virus) over the network by Kimura et al. (2009), who proposed blocking links approach for IC model to minimize the spread of unwanted information (virus) over the network. Authors calculated the contamination degree for the graph that represented average contamination degree for all graph nodes and edges after removing the chosen number of edges. Edge bond percolation method was used to estimate the number of blocked links, by computing the number of strongly connected components for a specific node '*u*' beside the number of reachable nodes for the same node '*u*'. They used two data sets. The first data set was employed by tracing up to ten steps back in the trackbacks from the blog of the theme JR Fukuchiyama Line Derailment

Collision from google website. This network was represented in a digraph of 12,047 nodes and 79,920 links. The second data set was a list of people within Japanese Wikipedia. The total numbers of nodes and directed links were 9,481 and 245,044, respectively. They concluded that blocking the links between the nodes with high out-degrees is not certainly effective.

Besides Kuhlman et al. (2013) focussed on the Linear Threshold model. They considered a control by block edges strategy on weighted and non-weighted social networks. This work differentiates between simple (interacts with one node) and complex (interacts with at least two nodes) contagions. They solved the optimization problem numerically by transforming the problem into a random walk in the network and numerically running it. The control cost was their objective function considering the condition that the fraction of votes in favor of the controller's opinion must be at least a fixed threshold. The authors concluded that the edge blocking method was effective in simple and complex contagions, for both non-weighted and weighted graphs.

Tong et al. (2012) used edge insertions and deletions for maximizing and minimizing the cascade. For the minimizing problem, they used the edge immunization approach by selecting a set k of edges whose deletion leads to grate drop in the first eigenvalue that leads to dissemination of the propagation process. Tong et al. used the approach input as a graph represented by '$N \times N$' matrix and a budget 'k', representing a set of edges that, if deleted, would lead to a drop in propagation. Firstly the algorithm makes certain that all eigen scores are non-negative. Then the left and right eigen scores for each edge are calculated that return the highest edges score. The authors simulate the diffusion via susceptible-infected-susceptible 'SIS' diffusion model on several types of network and compared the edge deletion with other approaches, like select edge randomly, select the highest degree edges, select highest page rank links and select the highest sign score edges. Tong et al. illustrated that their approach always led to the greatest reduction in eigenvalue and was scalable for large graphs.

Influence Minimization via Node Removal

Callaway et al. (2000) applied the percolation model as the epidemic behavior of the network. In the percolation model, as in the spread of computer virus, the node gets infected if it is substituted for infection. They allowed removal of vertices from the network in an order depending on their degree. They studied the network resilience via two techniques—firstly: removing random selected nodes or links; secondly: targeting deletion of network nodes or links. The network follows the random poison degree destruction. These networks are unlike the real world networks that frequently follow the power law degree disruption.

Increasing the random removal of real world network nodes does not damage the network connectivity as removing the hubs. Results showed that removing the higher degree ranked nodes damaged the Nestor connectivity effectively.

Albert and collaborators (2000) reported that the network was said to be robust if it maintains its functions even if some of its nodes fail or stop interacting. The study focussed only on the topological aspects of robustness caused by edge node removal. The results reported that scale-free networks are more robust than random networks against random node failures. On the other hand, such networks are more vulnerable and bad in information cascade when the most connected nodes are targeted since the graph connectivity has been damaged.

Wang et al. (2013) argued that removing nodes in order to decrease the out-degrees worked well in preventing the spread of negative information. Authors minimize the negative influence in the social networks by finding a good approximate solution via the greedy algorithm. They adopted Independent Cascade (IC) model for their solution. Results showed that minimizing the spread of negative information by blocking nodes in social networks was effective and proposed a greedy method for effective compared with classical centrality-based baseline methods.

Limiting Information Spread via Control Strategy

Given a negative active status, it was found that '*k*' positive seeds minimize the further negative influence and seeds maximize the number of blocked nodes from being activated by rumor cascade.

As an extension for IC model, Budak et al. (2011) studied the competing cascade on IC model and showed that for the general extension of the IC model in which good influence and bad influence have a separate set of parameters, the influence function is non-negative, is submodular only if the good influence probability is restricted to '1' or has the same probability as the negative influence. In addition, they reported that the influence function is monotone and that limiting bad information is NP-hard problem and provided a greedy approximate solution. The greedy solution for competitive cascade compared with using Monte Carlo simulation with classical heuristic centrality measures showed effectiveness. On the other hand, He et al. (2012) argued that misinformation always won because people are more likely to believe in negative opinions. They studied influence-blocking maximization in social networks under the competitive Linear Threshold model and proved submodularity of the objective function under that model.

Tsai et al. (2012) used game modelling for controlling the contagion. They started with the attacker that selected negative nodes to maximize the negative cascade while the defender selected positive nodes that minimize the attacker influence. The contagion can only take place once along any network edge and once a node is affected, it stays affected during the diffusion process. The model maximization strategy uses Nash equilibrium where both move simultaneously.

Kuhlman et al. (2013) addressed a control strategy by using the voter model in combination with Independent Cascade model (IC). They assumed that two opinions are propagated over the network. In addition, they formalized a new parameter called minimum opinion control factor (MOCF) to control opinions over a network in which the highest-degree nodes are controlled. They transformed the problem into an optimization problem, Given a graph G and a 'k' number of frozen nodes on a specific opinion, the objective was to reach the desired threshold, called the minimum average opinion. The average opinion is calculated for the entire graph depending on all the nodes opinions. The method was applied to several synthetic and real graphs and the authors found that MOCF depended on the network structure.

In addition, Yildiz et al. (2012) extended the voter model by examining the binary opinion where the nodes were classified into two groups—the leaders and followers. The leader nodes had a fixed state and did not change their opinion nor were influenced by neighbors opinions but they influenced others. The results led to a consensus and the node adopted the majority opinion from among its neighbors.

Ragia et al. (Ibrahim et al. 2017) controlled the rumor cascade by selecting the initial set of 'k' nodes to be immunized and to prevent the cascade from further spread. They used the CELF optimized algorithm to select a number of 'k' nodes for parallel processing. Firstly, the digraph was infected by a rumour, but after delay the 'd' rumor cascade was caught. Immunization algorithm selected set 'S' of chosen nodes to be

---

Algorithm 4. Greedy Immunization

Infected digraph (G) and k: number of initial nodes

  Output: selected seed set S

    initialization $S = \phi$, R: set of rumor nodes

FOR i := 1 to k

FOR $v \in V \backslash R \cup S_i$

$$\Delta_v = \sigma(V \backslash R \cup S_i) - (\sigma(V \backslash R \cup S_i) \cup \sigma(v))$$

$S_i = S_i \cup \text{argmax}_{v \in V \backslash R \, Si} \Delta_v{}^*$

ENDFOR

ENDFOR

  return(S)

---

immunized. The Δ_v function returned the average marginal gain for each node 'v'. The marginal gain was calculated on each iteration for a number of simulation times #NO$_{sim}$. For each 'i' in 'k', the node with the highest marginal gain Δ_v^* was added to set 'S' of the chosen seed nodes.

Applications and Challenges

Implementation of information cascade over social networks is rarely available and even if it's available, it should be re implemented in the different languages or formats that would be proper for the conducted data set. Although re-implementing the existing contributions needs high-level qualified facilities that could speed up the process, the unavailability and required facilities make it very difficult to evaluate the various contributions and compare them. In addition, since the invention of IC and LT models for information cascade in 2002, these models do not examine in reality or in real benchmark data set. There is a deep need to implement the model that learns from data and updates the results regarding the real ongoing cascade.

References

Albert, R., H. Jeong and A. L. Barabási. (2000). Error and attack tolerance of complex networks. Nature 406: 378–382.

BBC. (December 2016). False rumours and fake news cloud India's currency plan, Retrieved from http://www.bbc.com/news/blogs-trending-38318545.

Budak, C., D. Agrawal and A. El Abbadi. (2011, March). Limiting the spread of misinformation in social networks. pp. 665–674. *In*: Proceedings of the 20th International Conference on World Wide Web. New York, NY, USA.

Business Insider. (March 2013). Oreo's Super Bowl Power-Outage Tweet Was 18 Months In The Making, Retrieved from http://www.businessinsider.com/oreos-super-bowl-power-outage-tweet-was-18-months-in-the-making-2013-3.

Callaway, D. S., M. E. Newman, S. H. Strogatz and D. J. Watts. (2000). Network robustness and fragility: Percolation on random graphs. Physical Review Letters 85: 54–68.

Centola, D. and M. Macy. (2007). Complex contagions and the weakness of long ties. American Journal of Sociology 113: 702–734.

Chen, W., C. Wang and Y. Wang. (2010, July). Scalable influence maximization for prevalent viral marketing in large-scale social networks. pp. 1029–1038. *In*: Proceedings of the 16th ACM SIGKDD International Conference on Knowledge Discovery and Data Mining. Washington, DC, USA.

CNBC. (September 2015). Facebook are not going to charge, Retrieved from http://www.cnbc.com/2015/09/29/no-facebook-is-not-going-to-charge-users-a-privacy-fee.html.

Cohen, E. (1997). Size-estimation framework with applications to transitive closure and reachability. Journal of Computer and System Sciences 55: 441–453.

Coleman, J. S., E. Katz and H. Menzel. (1966). Medical Innovation: A Diffusion Study. Bobbs-Merrill Co.

Coleman, J. S., E. Katz and H. Menzel. (1966). Medical Innovation: A Diffusion Study. New York. Bobbs-Merrill 8: 15–24.

Coleman, J. S., E. Katz and H. Menzel. (1966). Medical Innovation: A Diffusion Study. New York: Bobbs-Merrill. USA.

Easley, D. and J. Kleinberg. (2010). Networks, Crowds, and Markets: Reasoning about a Highly Connected World. Cambridge University Press.

Foxnews. (March 2014). Catching up with Ted 'Golden Voice' Williams, Retrieved from http://www.foxnews.com/entertainment/2014/03/14/catching-up-with-ted-golden-voice-williams.html.

Goyal, A., W. Lu and L. V. Lakshmanan. (2011, December). Simpath: An efficient algorithm for influence maximization under the linear threshold model. *In*: Data Mining (ICDM), 2011 IEEE 11th International Conference on. pp. 211–220.

Goyal, A., W. Lu and L. V. Lakshmanan. (2011, March). Celf++: optimizing the greedy algorithm for influence maximization in social networks. pp. 47–48. *In*: Proceedings of the 20th International Conference Companion on World Wide Web. Hyderabad, India.

Griliches, Z. (1957). Hybrid corn: An exploration in the economics of technological change. Econometrica, Journal of the Econometric Society 25: 501–522.

Guille, A., H. Hacid, C. Favre and D. A. Zighed. (2013). Information diffusion in online social networks: A survey. ACM Sigmod Record 42: 17–28.

He, X., G. Song, W. Chen and Q. Jiang. (2012, April). Influence blocking maximization in social networks under the competitive linear threshold model. pp. 463–474. *In*: Proceedings of the 2012 SIAM International Conference on Data Mining. Society for Industrial and Applied Mathematics.

Ibrahim, R. A., H. A. Hefny and A. E. Hassanien. (2016, October). Controlling rumor cascade over social networks. pp. 456–466. *In*: International Conference on Advanced Intelligent Systems and Informatics Springer International Publishing. Egypt.

Ibrahim, R. A., H. A. Hefny and A. E. Hassanien. (2016, October). Group impact: Local influence maximization in social networks. pp. 447–455. *In*: International Conference on Advanced Intelligent Systems and Informatics. Springer International Publishing. Egypt.

Jiang, Q., G. Song, G. Cong, Y. Wang, W. Si and K. Xie. (2011, August). Simulated annealing based influence maximization in social networks. *In*: AAAI 11: 127–132.

Kempe, D., J. Kleinberg and É. Tardos. (2003, August). Maximizing the spread of influence through a social network. pp. 137–146. *In*: Proceedings of the Ninth ACM SIGKDD International Conference on Knowledge Discovery and Data Mining. ACM. Washington, D.C., USA.

Kimura, M., K. Saito and H. Motoda. (2009). Blocking links to minimize contamination spread in a social network. ACM Transactions on Knowledge Discovery from Data (TKDD), New York, NY, USA. 9: 1–9:23.

Kleinberg, J. (2007). Cascading behavior in networks: Algorithmic and economic issues. Algorithmic Game Theory 24: 613–632.

Krackhardt, D. (1987). Cognitive social structures. Social Networks 9: 109–134.

Kuhlman, C. J., V. A. Kumar and S. S. Ravi. (2013). Controlling opinion propagation in online networks. Computer Networks 57: 2121–2132.

Leskovec, J., A. Krause, C. Guestrin, C. Faloutsos, J. VanBriesen and N. Glance. (2007, August). Cost-effective outbreak detection in networks. pp. 420–429. *In*: Proceedings of the 13th ACM SIGKDD International Conference on Knowledge Discovery and Data Mining. San Jose, California, USA.

Li, H., S. S. Bhowmick, A. Sun and J. Cui. (2015). Conformity-aware influence maximization in online social networks. The VLDB Journal 24: 117–141.

Mossel, E. and S. Roch. (2007, June). On the submodularity of influence in social networks. pp. 128–134. *In*: Proceedings of the Thirty-ninth Annual ACM Symposium on Theory of Computing, San Diego, California, USA.

Nemhauser, G. L., L. A. Wolsey and M. L. Fisher. (1978). An analysis of approximations for maximizing submodular set functions - I. Math. Program 14: 265–294.

Peary, B. D., R. Shaw and Y. Takeuchi. (2012). Utilization of social media in the east Japan earthquake and tsunami and its effectiveness. Journal of Natural Disaster Science 34: 3–18.

Pedro Domingos and Matt Richardson. (2001). Mining the network value of customers. pp. 57–66. *In*: Proceedings of the Seventh ACM SIGKDD International Conference on Knowledge Discovery and Data Mining (KDD '01). ACM, New York, NY, USA.

Richardson, M. and P. Domingos. (2002, July). Mining knowledge-sharing sites for viral marketing. pp. 61–70. *In*: Proceedings of the Eighth ACM SIGKDD International Conference on Knowledge Discovery and Data Mining. Alberta, Canada.

Ryan, B. and N. Gross. (1943). The diffusion of hybrid seed corn in two Iowa communities. Rural Sociology 8: 15–24.

Sakaki, T., F. Toriumi and Y. Matsuo. (2011, December). Tweet trend analysis in an emergency situation. *In*: Proceedings of the Special Workshop on Internet and Disasters (p. 3).

The Guardian. (March 2011). Japan earthquake: Aid flows in from across the world, Retrieved from https://www.theguardian.com/world/blog/2011/mar/11/japan-earthquake.

Synthesio. (December 2013). Justin Bieber Nets the Most Death Rumors on Twitter, Retrieved from https://synthesio.com/corporate/newsroom/justin-bieber-nets-death-rumors-twitter-report.

The CNN. (January 2012). Conflicting reports about whether Mubarak has died, Retrieved from http://edition.cnn.com/2012/06/19/world/meast/egypt-mubarak/.

The Independent. (June 2011(. ISIS: Fake propaganda statement prompts false reports of leader Abu Bakr al-Baghdadi's 'death in US air strikes', Retrieved from http://www.independent.co.uk/news/world/middle-east/isis-fake-propaganda-statement-sparks-false-reports-of-leader-abu-bakr-al-baghdadis-death-in-us-air-a7081576.html.

THEEVERGE. (October 2014). Fake-News, Retrieved from https://www.theverge.com/2014/10/22/7028983/fake-news-sites-are-using-facebook-to-spread-ebola-panic.

Tong, H., B. A. Prakash, T. Eliassi-Rad, M. Faloutsos and C. Faloutsos. (2012, October). Gelling, and melting, large graphs by edge manipulation. pp. 245–254. *In*: Proceedings of the 21st ACM International Conference on Information and Knowledge Management.

Tsai, J., T. H. Nguyen and M. Tambe. (2012, July). Security games for controlling contagion. *In*: Proceedings of the Twenty-Sixth AAAI Conference on Artificial Intelligence. Toronto, Ontario, Canada.

Wang, S., X. Zhao, Y. Chen, Z. Li, K. Zhang and J. Xia. (2013, January). Negative influence minimizing by blocking nodes in social networks. pp. 134–136. *In*: AAAI (Late-Breaking Developments).

Wang, Y., G. Cong, G. Song and K. Xie. (2010, July). Community-based greedy algorithm for mining top-k influential nodes in mobile social networks. pp. 1039–1048. *In*: Proceedings of the 16th ACM SIGKDD International Conference on Knowledge Discovery and Data Mining.

Yildiz, E., A. Ozdaglar, D. Acemoglu, A. Saberi and A. Scaglione. (2013). Binary opinion dynamics with stubborn agents. ACM Transactions on Economics and Computation 1: 19.

Zafarani, R., M. A. Abbasi and H. Liu. (2014). Social Media Mining: An Introduction. Cambridge University Press. UK.

CHAPTER **10**

Swarm-based Analysis for Community Detection in Complex Networks

Khaled Ahmed,[1,*,#] *Ramadan Babers,*[2,*] *Ashraf Darwish*[2,*] and *Aboul Ella Hassanien*[1,*]

Introduction

Social network is a rapid communication method that allows thousands or millions of persons (users) to interact and exchange information through social network (Kramer et al. 2014). Online users spend their time on these networks, sharing information. There are different types of social networks based on the usage of these networks, such as gaming, online journals reading, dating, learning, music sharing, video sharing, micro-blogger and learning networks (Monica-Ariana and Anamaria-Mirabela 2014). Social network (SN) is a graph Z that is basically based on a set of nodes (N) and a set of edges (E), and is presented in Equation 1. These nodes represent the online persons or users while the interactions and communications between these users are edges (Crnovrsanin et al. 2014).

$$Z = N + E \tag{1}$$

[1] Faculty of Computers and Information, Cairo University, Egypt.
[2] Faculty of Science, Helwan University, Egypt.
* Scientific Research Group in Egypt; www.egyptscience.net
Emails: Ramadanfm@gmail.com; ashraf.darwish.eg@ieee.org; aboitcairo@gmail.com
Corresponding author: Khaled.elahmed@gmail.com

Social Network Analysis (SNA) is an urgent and valuable research trend based on understanding and mining the social networks' data in order to get valuable insights for helping decision makers in different domains, such as social domain in an election (Dheer and Verdegem 2014), quality of education (Helou and Rahim 2014) and public opinions and interests (Anstead and O'Loughlin 2015), medical domain in diseases health map (Widener and Li 2014), collecting patient experience (Greaves et al. 2013) and industrial domain in online advertising, delivering services and products to the interested or targeted persons (Ravi and Ravi 2015).

Community Detection (CD) is social network analysis method which some researchers consider as a prepossessing phase before any type of social network analysis is undertaken and other researchers consider the more accurate community detection more valuable and more insightful from social networks. It helps in dividing the social networks into a dynamic number of clusters or communities, with each cluster or community consisting of a set of nodes that have connective edges in the network. Each resultant cluster has nodes with the same interest or objective function, such as authors that citep the same journal, online users with the same interests and employees with the same online opinion. Figure 1 presents the structure of the basic social network with its nodes and edges (Maivizhi et al. 2016).

Meta heuristics algorithms (swarm optimization algorithms) help in optimizing and improving the accuracy and performance of many research techniques and approaches, such as feature selection, machine learning, image processing that helps in different domains and trends, such as medical, industrial and social domain (Hassanien and Emary 2016). Researchers present new approaches for community detection problem-based swarm optimization algorithms, such as discrete Krill herd (Ahmed et al. 2015), discrete bat swarm (Hassan et al. 2015b), artificial fish swarm (Hassan et al. 2015a), cuckoo search (Babers and Hassanien 2017), chicken swarm (Hassan et al. 2015b), ant loin (Babers et al. 2015a) and loin

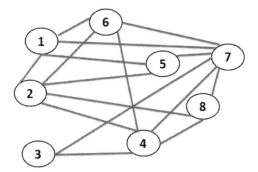

Fig. 1. Social network as a graph of nodes and edges.

algorithm (Babers et al. 2015b). This paper presents a comparative analysis based on experimental results for previous seven swarms for community detection using seven quality measures, such as NMI, modularity, ground truth, recall, precision, F-measure and accuracy over complex networks, such as Zachary Karate Club, Bottlenose Dolphin, American College football and Facebook data set.

Problem Definition

Community detection is an important prepossessing phase for more accurate social network analysis so that the more accurate community detection is the more accurate is the social network analysis. These enhanced results of SNA will reflect as benefits in many domains, such as sentiment analysis, online advertising, predication of public opinion and many other problems. It has many research challenges, such as:

- Finding a dynamic number of clusters for social network.
- Dividing the network into clusters with each cluster containing optimized set of nodes with high connectivity edges.
- Evaluating these clusters' quality by checking if the existing nodes are the best and the algorithm has to stop or the cluster needs to be reconstructed again and the algorithm containing search for a better solution.

Cuckoo Search Algorithm

Traditional Cuckoo Search Algorithm

Cuckoo Search Algorithm (CSA) was proposed by (Yang and Deb 2014). CSA is a meta heuristic search algorithm, which is inspired by cuckoo birds that have a unique behavior when searching for their nests (Yildiz 2013). The best nest, which contains the best solution (egg) will produce the second generation of solution. The probability of discovering the cuckoos' egg by the host bird is $p_a \in [0, 1]$. The technique used by the cuckoo for searching the suitable nest to lay its egg is Levy fight (Yang and Deb 2009).

Cuckoo Search Algorithm for Community Detection in Social Networks

Cuckoo search algorithm needs some modifications to fit in community detection problem in social networks. The egg existing in the nest is a solution and new one laid by the cuckoo represents a new solution. Parameters' values used in the algorithm are $I = 10$ (initial number of iterations), I_{max} = 100 (maximum number of iterations) and number of trails is 10, swarm size is initialized according to the nodes of the input data set, nest positions are initialized randomly, while eggs are calculated based on fitness values, step length s = 1000 and step size α = 1.5 (Babers and Hassanien 2017).

Figure 2 shows the steps of cuckoo search algorithm to detect communities in social network.

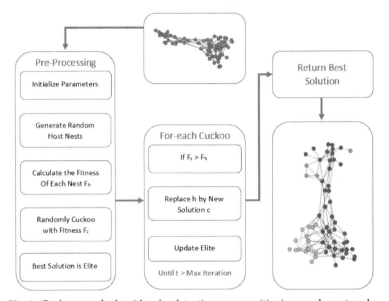

Fig. 2. Cuckoo search algorithm for detecting communities in complex networks.

Lion Swarm Algorithm

Traditional Lion Algorithm

Wang et al. (2012) and Yazdani and Jolai (2016) present Lion Optimization Algorithm (LOA), which is based on the lions' social behaviors in nature. The lion is the strongest mammal in the world as it has a unique social behavior which keeps only the stronger lion in herd in every generation.

Lion Algorithm for Community Detection in Social Networks

Lion Optimization Algorithm (LOA) is redesigned to fit complex networks cluster with community detection. LOA has three main steps to detect communities in social networks, such as initialization, mating and territorial defense. The initialization phase is used to generate a random population over the solution space. Every solution in LOA is called lion that is represented by:

$$Lion = [x_1, x_2, \cdots, x_{Nvar}] \tag{2}$$

where N_{var} is dimensional space. The ratio of nomad lions is %N generated randomly and the rest of the population is resident lions. Mating—by this step, the new best solutions are derived from the existing one. In this step the lion kills the weak cubs to ensure that the new solutions are best. Territorial defense—in this phase, the comparison between fitness of lions in the pride and nomadic lions are held. If the fitness of nomadic lion is better than the resident one replacing process occurs. LOA's parameters for detecting communities in complex network are $N = 0.2$, $R = 0.8$, age mat = 3, distance = 0.4 and step = 0.05 n.

Figure 3 shows the steps of algorithm to detect communities in social network Babers et al. (2015b).

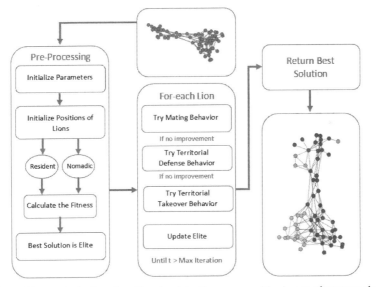

Fig. 3. Lion optimization algorithm for detecting communities in complex networks.

Ant Lion Swarm Algorithm

Traditional Ant Lion Algorithm

The ant lion is an insect which has a special behavior in catching its preys by digging a trap in sand as a cone-shaped pit in different sizes depending on the level of hunger and shape of the moon. The Ant Lion Optimizer (ALO), was presented by Mirjalili (2015). ALO is simulating the hunting behavior of ant lions. Since ants move stochastically and randomly in nature when searching for food, a random walk is chosen for modeling ants' movements.

Ant Lion Algorithm for Community Detection in Social Networks

Ant lions update their positions in algorithm with random walk at every iteration as shown before in LOA. The position of ant lion represented by x consists of n elements $(x_1, x_2, \cdots, x_n)$. This position represents a solution in the search space and the fitness value of each ant lion represents the amount of ants that are ant lion can prey on. The number of ants can find in the same position of ant lion is expressed as $y_i = f(x_i)$, where y_i is the objective function value assigned to x_i. Genetic algorithm Pizzuti (2008) represents the membership between individuals and simulates the movements in the solution x_i toward a solution x_j by using crossover. ALO's parameters are distance = 0.4 and step = 0.05 n. Figure 4 shows the steps of algorithm to detect communities in social network (Babers et al. 2015b).

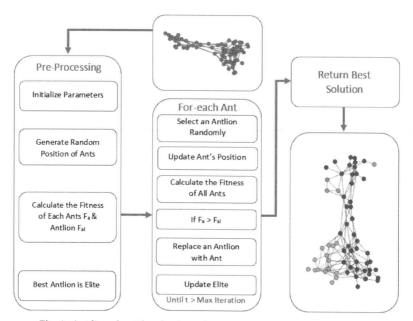

Fig. 4. Ant lion algorithm for detecting communities in social networks.

Chicken Swarm Algorithm

Traditional Chicken Swarm Algorithm

Chicken swarm optimization (CSO) algorithm (Meng et al. 2014) is a new optimization algorithm which simulates the movements of chicken in farms or in nature. Chicken movements have a structure while searching for food (objective), which start with rooster as leader. Then two hens and the rest are chicks which follow their mothers and leader (rooster). Chicken swarm is divided into groups with respect to the previous structure. Each group has one rooster, which has the best fitness value, while chicks have the worst value and the rest are hens.

Chicken Swarm Algorithm for Community Detection in Complex Social Networks

Chicken swarm algorithm (ACSO) for community detection (Ahmed et al. 2016) is illustrated in Fig. 5 with parameters' values of maximum iteration number as 300, minimum iteration number as 50, and the position of rooster, hens and chicks are started randomly. The food position is based on fitness function, $W = 0.6$, $FL = 0.7$, $C = 0.4$, swarm size n consists of 0.78 chicks, 0.04 hens, 0.03 chicks-mother and 0.15 roosters.

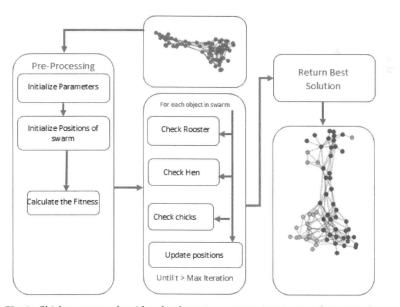

Fig. 5. Chicken swarm algorithm for detecting communities in complex networks.

Krill Herd Algorithm

Basic Krill Herd Algorithm

Krill herd algorithm is an optimization algorithm (Gandomi and Alavi 2012, Saremi et al. 2014), which presents the movements of individual krill's in nature in search for food. Krill herd algorithm consists of three steps, such as induced motion (N), foraging motion (F) and random diffusion (D).

Krill Herd Algorithm for Community Detection in Complex Social Networks

Krill herd algorithm (AKHO) for community detection (Ahmed et al. 2015) is presented in Fig. 6 while the used parameters' values are swarm size n which is calculated based on input data set nodes, iteration number (Imax) = 300, initial or minimum number of iteration = 50.

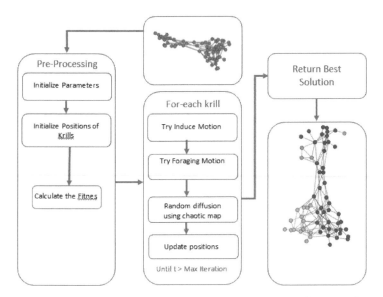

Fig. 6. Krill herd algorithm for detecting communities in complex networks.

Fish Swarm Algorithm

Traditional Fish Swarm Algorithm

Fish swarm algorithm is a nature-inspired optimization algorithm (Li et al. 2002), which simulates the behavior of fish swarm in nature. Fish swarm moves together in groups searching for food and to prevent themselves from any enemy. Fish swarm achieves high results of performance and accuracy in solving many research problems, such as feature selection (Lin et al. 2015), clustering and routing (Yazdani et al. 2013).

Fish changes its position based on its vision. If the fish finds new interesting position for food, it will follow this position, which is called 'visual position', else the fish goes in its path to explore the space until finding a better and more interesting position.

Fish swarm movements are divided into:

- *Praying*: Fish is moving towards food in the water. It is trying to reach food in the water with a number of trials. After this number of trials, if the fish is not able to find food, then it moves a step randomly in the search space.

- *Swarming*: Fish swim in groups to protect themselves from their enemy. So in this step, the distance between the fish's position and each group center is calculated. The group with the smallest distance is joined by the fish and swim towards in the search space.

- *Following*: When one partner of the fish group finds food in the water, the fish in the same group will follow to reach this position and update their current positions.

- *Leaping*: Fish stop searching somewhere in the search pace and start a new random step to get a better solution. If the fish can't achieve a better solution, then the experiment is restarted.

Artificial Fish for Community Detection in Social Networks

Artificial fish algorithm (AFSA) for community detection (Hassan et al. 2015a) is stated in Fig. 7 while algorithm parameters' values of swarm size n is calculated based on the input data sets nodes, iteration number (Imax) = 300, initial or minimum number of iteration = 5, step is 0.2 size of swarm, visual = 0.8, crowd factor is 0.3 and preying check = 10.

Bat Swarm Algorithm

Traditional Bat Swarm Algorithm

Bat swarm optimization algorithm (Yang 2010) is a nature-inspired optimization algorithm, which describes how bats act in nature. Each bat emits ultrasound pulses for detecting, exploring the surrounding object and hunting as well. Individual bats send ultrasound pluses with high loudness and low frequency to explore a big area and to detect the

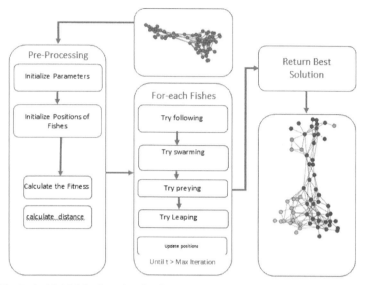

Fig. 7. Artificial fish algorithm for detecting communities in complex networks.

surrounding objects. If the bat detects the prey position, it will send more ultrasound pluses with high frequency and low sound.

Each bat has four features to change its position in the search space, while finding the best solution (prey), such as the emitted pulses frequency, loudness, velocity and its current position.

Artificial Bat for Community Detection in Social Networks

Artificial bat for community detection in complex social networks (Hassan et al. 2015b) is presented in Fig. 8, the values of the parameters are number of VB in the population $np = 100$ and the maximum number of iterations = 300 and minimum number of iteration is 50 and loudness is 0.6.

Data Sets and Quality Measures

Experiments are executed over four popular benchmarks—data sets, such as Zachary Karate Club, Bottlenose Dolphin, American College football and Facebook data set. Table 1 states the benchmarks networks' characteristics (Girvan and Newman 2002, Lusseau 2003, Leskovec and Mcauley 2012, Zachary 1977).

Experiments' quality measures are used to evaluate experiments results, such as NMI, modularity, ground truth, recall, precision, accuracy and F-measure (Hric et al. 2014, Harenberg et al. 2014).

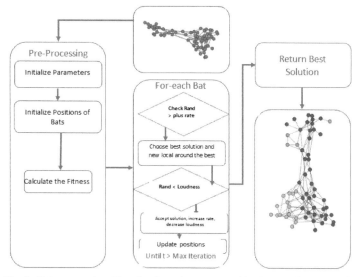

Fig. 8. Bat swarm algorithm for detecting communities in complex networks.

Table 1. Datasets characteristics.

Name	Nodes	Edges	Ground Truth
Zachary Karate Club	34	78	0.42100
Bottlenose Dolphin	62	159	0.39400
American College football	115	613	0.56300
Facebook	3959	84243	0.72300

Normalized Mutual Information (NMI) defines the ratio between the output cluster and predefined value (ground truth) for the network, while modularity measures the quality of the internal nodes in each cluster and recall, precision, F-measure and accuracy are used to measure the quality of the algorithm which are illustrated in Equations 3,4,5,6.

$$Recall = \frac{relevant \cap retrieved}{relevant} \tag{3}$$

$$Precision = \frac{relevant \cap retrieved}{retrieved} \tag{4}$$

$$F - Measure = \frac{2 * Recall * Precision}{Recall + Precision} \tag{5}$$

$$Accuracy = \frac{VR + RV_n}{VR + VR_n + RV_n + R_n V_n}$$ (6)

Where V is relevant, R is retrieved, V_n is not relevant and R_n is not retrieved.

Discussion and Experimental Analysis

Experiments are executed on a computer with 8 GB Ram, Core i7. Experiments are executed based on seven swarm optimization algorithms for community detection in complex social networks, such as a discrete krill herd, discrete Bat, Artificial fish swarm, cuckoo search, adaptive chicken swarm, ant loin and loin algorithm over four benchmark data sets, such as Zachary Karate Club, Bottlenose Dolphin, American College football and Facebook. Experimental results are recorded for an average of 19 runs for each algorithm over each data set. Each benchmark data set has predefined ground truth value to measure the algorithm's quality. Locus-based adjacency scheme is used for encoding and decoding tasks.

Experimental results for the seven algorithms over benchmark datasets are presented in Figs. 9, 10, 11 and 12. Table 2 presents results over Facebook dataset, Table 3 shows results over American College football dataset, Table 4 over Bottlenose Dolphin data set and Table 5 over Zachary Karate Club dataset. Experimental comparisons are executed for the results. Figure 9 presents results over Facebook dataset, Fig. 10 over American College football dataset. Figure 11 over Bottlenose Dolphin dataset and Fig. 12 over Zachary Karate Club data set.

The higher NMI and modularity against ground truth value reflect better quality community detection and structure, while the lower NMI and modularity values against ground truth reflect lower quality of the detected communities and of inner nodes. Recall and precision values state the ratio and percentage of the accurate and correct retrieved communities against ground truth, while the higher F-measure and accuracy state higher quality and performance of the model in the domain of community detection. Experimental results show that the swarm algorithms achieve same results and quality measures for small size datasets, while they have a small difference in results for medium size dataset. In big datasets, every swarm algorithm achieves different results and quality measures for community detection in complex social networks based on the swarm algorithm's mechanism and how it acts.

From experiments, we propose an ordered list starts from the best swarm optimization algorithm till the smallest and lowest quality swarm optimization algorithm for community detection problem in complex social network in case of small, medium or large-scale dataset. The

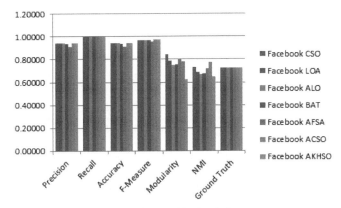

Fig. 9. Results over Facebook dataset.

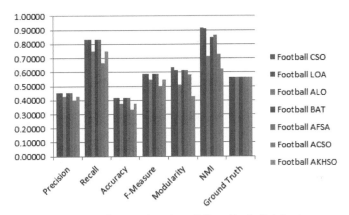

Fig. 10. Results over American College Football dataset.

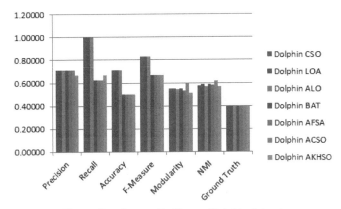

Fig. 11. Results over Bottlenose Dolphin dataset.

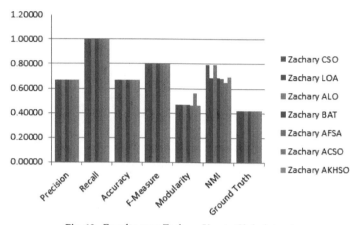

Fig. 12. Results over Zachary Karate Club dataset.

Table 2. Results over Facebook dataset.

	Algo	Precision	Recall	Accuarcy	F-measure	NMI	Modularity	Ground Truth
Facebook	CSO	0.94286	1.00000	0.94286	0.97059	0.84324	0.73000	0.723000
	LOA	0.94186	1.00000	0.94186	0.97006	0.78424	0.69000	0.72300
	ALO	0.94083	1.00000	0.94083	0.96951	0.75125	0.67000	0.72300
	BAT	0.93706	1.00000	0.93706	0.96751	0.75300	0.67374	0.72300
Data Set	AFSA	0.91176	1.00000	0.91176	0.95385	0.80500	0.72000	0.72300
	ACSO	0.94578	1.00000	0.94578	0.97214	0.78200	0.77400	0.72300
	AKHSO	0.94611	1.00000	0.94611	0.97231	0.62770	0.65120	0.72300

Table 3. Results over Football dataset.

	Algo	Precision	Recall	Accuracy	F-measure	NMI	Modularity	Ground Truth
Football	CSO	0.45455	0.83333	0.41667	0.58824	0.63520	0.91450	0.56300
	LOA	0.45455	0.83333	0.41667	0.58824	0.61320	0.90950	0.56300
	ALO	0.42857	0.75000	0.37500	0.54546	0.50980	0.71263	0.56300
	BAT	0.45455	0.83333	0.41667	0.58824	0.61200	0.84786	0.56300
Data Set	AFSA	0.45455	0.83333	0.41667	0.58824	0.61500	0.86500	0.56300
	ACSO	0.40000	0.66667	0.33333	0.50000	0.58200	0.73100	0.56300
	AKHSO	0.42857	0.75000	0.37500	0.54545	0.42520	0.62460	0.56300

Table 4. Results over Dolphin dataset.

	Algo	Precision	Recall	Accuarcy	F-measure	NMI	Modularity	Ground Truth
Dolphin	CSO	0.71429	1.00000	0.71429	0.83333	0.54920	0.57760	0.39400
	LOA	0.71429	1.00000	0.71429	0.83333	0.54910	0.58820	0.39400
	ALO	0.71429	1.00000	0.71429	0.83333	0.54211	0.57152	0.39400
	BAT	0.71429	0.62500	0.50000	0.66667	0.54980	0.58670	0.39400
Data Set	AFSA	0.71429	0.62500	0.50000	0.66667	0.53000	0.58000	0.39400
	ACSO	0.71429	0.62500	0.50000	0.66667	0.59200	0.62100	0.39400
	AKHSO	0.66667	0.66667	0.50000	0.66667	0.51230	0.56590	0.39400

Table 5. Results over Zachary dataset.

	Algo	Precision	Recall	Accuarcy	F-measure	NMI	Modularity	Ground Truth
Zachary	CSO	0.66660	1.00000	0.66670	0.80000	0.47020	0.79370	0.42100
	LOA	0.66660	1.00000	0.66670	0.80000	0.46960	0.68730	0.42100
	ALO	0.66660	1.00000	0.66670	0.80000	0.46960	0.79370	0.42100
	BAT	0.66660	1.00000	0.66670	0.80000	0.46960	0.68730	0.42100
Data Set	AFSA	0.66660	1.00000	0.66670	0.80000	0.46000	0.68000	0.42100
	ACSO	0.66660	1.00000	0.66670	0.80000	0.56300	0.65200	0.42100
	AKHSO	0.66660	1.00000	0.66670	0.80000	0.46150	0.69370	0.42100

ordered list is cuckoo search algorithm, loin algorithm, adaptive chicken swarm, artificial fish swarm, artificial bat, ant loin algorithm and finally krill herd swarm algorithm. The swarm algorithm with better community detection results causes its start with solution, removes the worst solution, updates its structure until the best solution is obtained or is close to best and updates the whole space while swarm with good results causes the effect of random variable in search space in the swarm, such as in krill state with induced motion, foraging, random diffusion.

Open Problems and Research Gaps

There are a lot of open research challenges for community detection in complex social networks for more accurate community detection which are illustrated as the following:

- Applying chaotic maps with swarm to improve community detection as the effect of the random variable in swarm algorithm.
- Use of quantum theory with swarm algorithms as a hybrid model for community detection.
- Parallel version of the swarm optimization algorithm for large-scale complex network.
- Hybrid two swarm algorithms which start swarm algorithm for optimizing the number of clusters and another swarm algorithm for optimizing the nodes in each cluster.
- Optimizing traditional community detection techniques by swarm algorithm as a hybrid model.
- Hybrid deep learning model with swarm algorithm.
- Information propagation in social communities.
- User's influence or nodes influence in social communities.
- Spam users in social community detection such as spam node or spam cluster.
- Improving sentiment analysis for online based on using community detection as prepossessing phase.
- Dividing the Internet of Things (IOT), objects, into Communities of Things (CIOT).

Conclusion and Future Works

This paper presents a comparative analysis based on seven swarm optimization algorithms, such as krill herd, bat swarm, artificial fish swarm, cuckoo search, chicken swarm, ant loin and loin algorithm for community detection over four benchmarks data sets, such as Zachary Karate Club, Bottlenose Dolphin, American College football and Facebook using seven quality measures of NMI, modularity, ground truth, recall, precision, F-measure and accuracy. The results are recorded with average 19 runs for each data set. Future works are applying more swarm algorithms and chaotic theory with swarm for community detection and community detection on internet of thing (IOT) to present new model 'CIOT'.

References

Ahmed, K., A. I. Hafez and A. E. Hassanien. (2015). A discrete krill herd optimization algorithm for community detection. In 2015 11th International Computer Engineering Conference (ICENCO), IEEE pp. 297–302.

Ahmed, K., A. E. Hassanien, E. Ezzat and P.-W. Tsai. (2016). An adaptive approach for community detection based on chicken swarm optimization algorithm. In International Conference on Genetic and Evolutionary Computing, Springer pp. 281–288.

Anstead, N. and B. O'Loughlin. (2015). Social media analysis and public opinion: The 2010 UK general election. Journal of Computer-Mediated Communication 20(2): 204–220.

Babers, R., N. I. Ghali, A. E. Hassanien and N. M. Madbouly. (2015a). Optimal community detection approach based on ant lion optimization. In 2015 11th International Computer Engineering Conference (ICENCO), IEEE pp. 284–289.

Babers, R., A. E. Hassanien and N. I. Ghali. (2015b). A nature-inspired metaheuristic lion optimization algorithm for community detection. In 2015 11th International Computer Engineering Conference (ICENCO), IEEE pp. 217–222.

Babers, R. and A. E. Hassanien. (2017). A nature-inspired metaheuristic cuckoo search algorithm for community detection in social networks. International Journal of Service Science, Management, Engineering, and Technology (IJSS-MET) 8(1): 50–62.

Crnovrsanin, T., C. W. Muelder, R. Faris, D. Felmlee and K.-L. Ma. (2014). Visualization techniques for categorical analysis of social networks with multiple edge sets. Social Networks 37: 56–64.

Dheer, E. and P. Verdegem. (2014). Conversations about the elections on twitter: Towards a structural understanding of twitters relation with the political and the media field. European Journal of Communication 29(6): 720–734.

Gandomi, A. H. and A. H. Alavi. (2012). Krill herd: A new bio-inspired optimization algorithm. Communications in Nonlinear Science and Numerical Simulation 17(12): 4831–4845.

Girvan, M. and M. E. Newman. (2002). Community structure in social and biological networks. Proceedings of the National Academy of Sciences 99(12): 7821–7826.

Greaves, F., D. Ramirez-Cano, C. Millett, A. Darzi and L. Donaldson. (2013). Use of sentiment analysis for capturing patient experience from free-text comments posted online. Journal of Medical Internet Research 15(11).

Harenberg, S., G. Bello, L. Gjeltema, S. Ranshous, J. Harlalka, R. Seay, K. Padmanabhan and N. Samatova. (2014). Community detection in large-scale networks: a survey and empirical evaluation. Wiley Interdisciplinary Reviews: Computational Statistics 6(6): 426–439.

Hassan, E. A., A. I. Hafez, A. E. Hassanien and A. A. Fahmy. (2015a). Community detection algorithm based on artificial fish swarm optimization. In Intelligent Systems' 2014, Springer pp. 509–521.

Hassan, E. A., A. I. Hafez, A. E. Hassanien and A. A. Fahmy. (2015b). A discrete bat algorithm for the community detection problem. In International Conference on Hybrid Artificial Intelligence Systems, Springer pp. 188–199.

Hassanien, A. E. and E. Emary. (2016). Swarm Intelligence: Principles, Advances, and Applications. CRC Press.

Helou, A. M. and N. Z. A. Rahim. (2014). The influence of social networking sites on students academic performance in malaysia. International Journal of Electronic Commerce 5(2): 247–254.

Hric, D., R. K. Darst and S. Fortunato. (2014). Community detection in networks: Structural communities versus ground truth. Physical Review E 90(6): 062805.

Kramer, A. D., J. E. Guillory and J. T. Hancock. (2014). Experimental evidence of massive-scale emotional contagion through social networks. Proceedings of the National Academy of Sciences 111(24): 8788–8790.

Leskovec, J. and J. J. Mcauley. (2012). Learning to discover social circles in ego networks. In Advances in Neural Information Processing Systems pp. 539–547.

Li, X.-l., Z.-j. Shao, J.-x. Qian et al. (2002). An optimizing method based on autonomous animats: Fish-swarm algorithm. System Engineering Theory and Practice 22(11): 32–38.

Lin, K.-C., S.-Y. Chen and J. C. Hung. (2015). Feature selection for support vector machines base on modified artificial fish swarm algorithm. In Ubiqui-tous Computing Application and Wireless Sensor, Springer pp. 297–304.

Lusseau, D. (2003). The emergent properties of a dolphin social network. Proceedings of the Royal Society of London B: Biological Sciences 270(Suppl 2): S186–S188.

Maivizhi, R., S. Sendhilkumar and G. Mahalakshmi. (2016). A survey of tools for community detection and mining in social networks. In Proceedings of the International Conference on Informatics and Analytics, ACM pp. 71.

Meng, X., Y. Liu, X. Gao and H. Zhang. (2014). A new bio-inspired algorithm: chicken swarm optimization. In International Conference in Swarm Intelligence, Springer pp. 86–94.

Mirjalili, S. (2015). The ant lion optimizer. Advances in Engineering Software 83: 80–98.

Monica-Ariana, S. and P. Anamaria-Mirabela. (2014). The impact of social media on vocabulary learning case study facebook. Annals of the University of Oradea, Economic Science Series 23(2): 120–130.

Pizzuti, C. (2008). Ga-net: A genetic algorithm for community detection in social networks. In Parallel Problem Solving from Nature–PPSN X, Springer pp. 1081–1090.

Ravi, K. and V. Ravi. (2015). A survey on opinion mining and sentiment analysis: tasks, approaches and applications. Knowledge-Based Systems 89: 14–46.

Saremi, S., S. M. Mirjalili and S. Mirjalili. (2014). Chaotic krill herd optimization algorithm. Procedia Technology 12: 180–185.

Wang, B., X. Jin and B. Cheng. (2012). Lion pride optimizer: An optimization algorithm inspired by lion pride behavior. Science China Information Sciences 55(10): 2369–2389.

Widener, M. J. and W. Li. (2014). Using geolocated twitter data to monitor the prevalence of healthy and unhealthy food references across the US. Applied Geography 54: 189–197.

Yang, X.-S. and S. Deb. (2009). Cuckoo search via levy flights. In Nature and Biologically Inspired Computing, 2009. NaBIC 2009. World Congress on, IEEE pp. 210–214.

Yang, X.-S. (2010). A new metaheuristic bat-inspired algorithm. Nature Inspired Cooperative Strategies for Optimization (NICSO 2010), pp. 65–74.

Yang, X.-S. and S. Deb. (2014). Cuckoo search: recent advances and applications. Neural Computing and Applications 24(1): 169–174.

Yazdani, D., B. Saman, A. Sepas-Moghaddam, F. Mohammad-Kazemi and M. R. Meybodi. (2013). A new algorithm based on improved artificial fish swarm algorithm for data clustering. International Journal of Artificial Intelligence 11(A13): 193–221.

Yazdani, M. and F. Jolai. (2016). Lion optimization algorithm (loa): a nature-inspired metaheuristic algorithm. Journal of Computational Design and Engineering 3(1): 24–36.

Yildiz, A. R. (2013). Cuckoo search algorithm for the selection of optimal machining parameters in milling operations. The International Journal of Advanced Manufacturing Technology 64(1-4): 55–61.

Zachary, W. W. (1977). An information flow model for conflict and fission in small groups. Journal of Anthropological Research 33(4): 452–473.

CHAPTER 11

Trustworthiness in the Social Internet of Things

B.K. Tripathy* and Deboleena Dutta

Introduction

Internet of Things (IoT) is a network of physical objects or things embedded with smart objects such as mobile robots, wireless sensors. They are the third wave of Internet which is supposed to have the potential to connect about 28 billion things by 2020 ranging from bracelets to cars. The term IoT, which was first proposed by Kevin Ashton, a British technologist, in 1999, has the potential to impact everything from new product opportunities to shop floor optimization to factory worker efficiency that will power top-line and bottom-line gains. It is believed that IoT will improve energy efficiency, remote monitoring and control of physical assets and productivity through applications as diverse as home security to condition monitoring on the factory floor. Now IoT are used in markets in the field of health care, home appliances and buildings, retail markets, energy and manufacturing companies, mobility and transportation, logistics companies and by media. IoT has millions of services, with strict real-time requirements and striking flexibility in connecting everyone and everything.

School of Computing Science and Engineering, VIT University, Vellore-632014, TN, India.
Email: deboleena.d@gmail.com
* Corresponding author: tripathybk@vit.ac.in

Social Internet of Things (SIoT) is a paradigm unifying IoT with social networks such that the human beings and devices interact with each other. It is obtained by integrating social networking concepts into the Internet of Things. It enables people and devices to interact, thus facilitating information sharing and enabling a variety of interesting applications.

IoT Architecture

Social Internet of Things is a system of devices that connect and interact directly with each other to send, receive, share or store meaningful and intelligent data through a specific layer known as Secure Service Layer (SSL), which helps in connecting to a control server and help in maintaining the data storage in the database or over the cloud. IoT can exist based on different types of heterogeneous devices. All devices must have the capability to sense data, capture data, store data temporarily and send data to other devices.

Initially, the internet of things was introduced with a three-layered architecture (Fig. 1). The layers consisted of Roman et al. (2011).

1. *Perception Layer*: Perception Layer is the layer which consisted of the RFID sensors used to sense all objects with RFID tagging for sending and receiving real-time information about the given object and its environment (Kosmats 2011).

2. *Network Layer*: This layer as the name suggests helps in forming the network around the objects to interact with each other. This creates a local layer of network around the object and within which interaction is possible.

3. *Application Layer*: This is the layer which is the user layer where applications show the actual real-time data. The data collected and stored from the lower layers was deployed in this layer for user to refer.

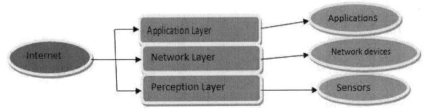

Fig. 1. Basic architecture of Internet of Things (three-layered architecture) (https://www. linkedin.com).

There are four types of communication in IoT system:

i. *Device-to-Device*: Unlike humans, devices or objects communicate with each other by forming a network locally around the objects. These devices interact without any existence of a middleware. Devices can share, receive send data or store them for future reference.

ii. *Device-to-User*: Devices can directly send and receive real time information with meaningful data to the user's phone or any other device with the help of the network.

iii. *Device-to-Server*: In this communication device the data is sent to server where the data is stored or it can use the cloud for storing big data for future reference.

iv. *Server-to-Devices*: The data stored in the server or cloud can be used by other devices when and wherever it is required; for instance, we can consider the doctor–patient real-time example to understand the above.

When a patient is suffers from a chronic illness, with the advent of new technology as Internet of Things, the patient need not be admitted in a hospital. He can stay at home and still be under observation by the doctor. Smart beds can provide the patient's movement details. With the help of a medical IoT band they can keep a track of the drugs which are being induced and the amount of drug injected can be maintained. Regular blood pressure, temperature, blood test, blood sugar, urine samples can be collected and tested. These reports are automatically sent to the doctor for reference. The stored data can be accessible to the patient's family or to the relevant doctor by the user's applications.

The architecture clearly describes how the total system works. A device takes the data from the human body and sends to the base station. From the base station, the data is stored in the server in cloud. From the server, the doctor can access the data. If required, even the patient can access the data by using applications in his mobile phone. Today, these interactions over the network can be accessed using the social network.

Social Networking

A social network comprises of a finite set of actors, who are social entities. These entities can be discrete individuals, corporate or collective social units. They are related to each other through some relations, establishing some linkage among them. Social network has grown in popularity as it enables researchers to study not only the social actors, but also their social relationships (Fiske 1992). Moreover, many important aspects of societal networks and their study lead to the study of behavioural science. Scientific study of network data can reveal important behaviour of the elements involved and social trends. It also provides insight into suitable changes in the social structure and roles of individuals in it. Important aspects of societal life are organised as networks. The importance of networks in society has put social network analysis at the forefront of social and behavioural science research. The presence of relational information is a critical and defining feature of the social network.

Today, we find that most people are attached to one or more social networks for one or more purposes like personal use, advertisement, business, to convey messages, etc. Social networks are applications (web-based) (Golbeck 2005, Golbeck et al. 2006, Golbeck 2008) which permit people to create a profile to share details with an association or provide information to clients. During the past few years, study on social networks has been a popular topic for research, leading to useful benefits for the users. They allow users to make new friends and to stay in contact with old ones, but also to make connections with people with similar interests and goals. Today, social networks provide and integrate an increasing number of communication and collaboration services as well as specific applications (Nitti 2014, Zhang et al. 2014).

With the help of social networking sites, not only students, professionals but also people from other parts of the society are actively engaged with social networking sites, e.g., Facebook, LinkedIn, Twitter, etc. Besides using these sites for personnel reasons, many researchers have also tried to analyze and measure the effect of this usage on society. However, most of them were analyzing, surveying or investigating for specific issues, such as its privacy, the user motive, and the user behaviour. It is a fact that the social networking site has already played a vitally important role in current professional and commercial online interactions. Despite all the surveys, the measurement of each of the issues rendering the analysis of impact, characteristic or motive of users (Nitti et al. 2012, Li et al. 2015).

Applications of SIoT

At present, social internet of things is the new emerging technology in possible fields that we can think of. These IP-enabled things (smart things) are capable of different ranges of communication, unlike IoT. SIoT has taken a different kind of momentum in recent years to be one of the most promising technologies in the near future. We can enable socially connected devices in the fields of engineering, medicine, smart homes, defence, health, etc. (https://www.linkedin.com, https://blog.apnic.net). Today internet of things extend from smart phone to smart homes to smart offices to smart shopping centers to smart cities. Things are getting broader and brighter with every passing day. We can find IoT devices that can be worn while doing our daily chores or can be installed in the office building.

The following are some of the applications of SIoT at present:

Smart Supermarket

Those were the days when we used to walk to our nearest grocery store with our grandfathers, wait with them while they spoke with the shopkeeper, got the required grocery in the bag and came home. Later this idea of grocery store was changed to super market. This is where we enter, start searching our grocery, carry them in a basket to the billing section–waiting in a queue till our chance, the cashier scans the barcode of each item, calculates the total, we pay the bill, collect the items in a bag and leave. Now, with the advent of IoT, Amazon has come up with the idea of no queues–collect the required grocery items from the racks and leave.

Amazon Go (Fig. 2) is a totally automated store; it is operated without any cashier at the billing section and nobody has to wait in the queue for their chance to come. The technology works in the store depend on smart phones and sensors are infused to detect the objects moving in the store. The concept is still new with only one pilot outlet set up in Seattle, USA (https://en.wikipedia.org).

Fig. 2. Amazon Go (Christianson and Harbison 1997).

Smart City

Smart city (Fig. 3) is the future of urbanization. It is the vision of Internet of Things where sensors, smartphones, actuators and smart objects connect and interact through the internet to gather real-time information of different 'objects', like weather forecast, parking space availability, society occupancy, keep a track of the people staying in the area, track pollution, water consumption, electricity consumption and so on.

Day-to-day check of an entire city can be managed by connecting digital objects to each other, using sensors, RFID and internet. To connect a city with such a vast area smartly, different networks like Home Area Network, Wide Area Network and Field Area Network are implemented and connected. Few examples of applications in such cities are smart homes, smart weather systems, smart energy consumption, pollution control, surveillance, and healthcare, etc.

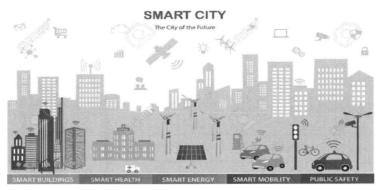

Fig. 3. A smart city.

Smart Healthcare

Medical IoT can be called the new era of a healthy life, where we can save time with the help of sensor devices. These devices in the future can minimize the need of visiting hospitals for health check-ups and not wait for prior appointments for any routine check-ups. The relationship of doctor-patient can still be maintained with the help of the connected devices over internet. There are times where middle-aged diabetic patients or high-blood-pressure patients need to maintain a record of their monthly sugar and blood pressure levels respectively. Those long queues to see the doctor or visit the hospitals to get a nurse to measure them can be easily avoided with the help of such devices. Medical is one of the most important fields where IoT has revolutionized by improving the lives

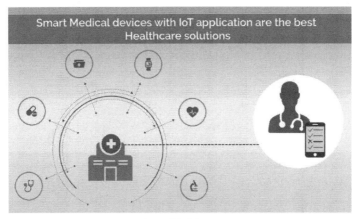

Fig. 4. Healthcare.

of many people around the world (Fig. 4). With every new invention in medical IoT, medical treatment has enhanced. Patients can be monitored remotely, discuss their problems directly with the doctors and doctors need not be there in person for the follow-up. Chronic patients can be monitored remotely, their drug doses can be automated. These devices sends alert to family members, doctors and nurses if any kind of signs are indicated which need to be checked upon (http://searchhealthit.techtarget.com). Nowadays doctors can get 3D-vision of the area of surgery, helping them to bypass those nerves or veins without touching them at all.

There are IoT-enabled laboratories for instant or routine pathological tests for patients. There are beds which can monitor sleep and movements of the patients. With the advent of sensor devices, life in a few years can be safe, healthy, easier and time-saving in terms of medicine.

Smart Home

Smart home is another area where there have been many new inventions, like Amazon Echo, Phillips Hue, Hive Active Heating 2, Netatmo Welcome, etc. These appliances help simplify our lives at our homes and offices (Fig. 5). There devices can detect sunlight and the changing color of glass to cool down the temperature inside the office floor. There are devices for home where it detects the temperature outside and starts the heater and the room temperature gadget. These devices and more are all automated. For security purpose, we have Dropcam, which is a surveillance camera with improved night vision, audio and video camera. This device can be connected to our mobile phones to indicate any unknown person entering

Fig. 5. Smart home (https://www.linkedin.com).

the boundary. We also have Chamberlain's garage-door opener, which can be used to remotely open and close the garage door. This is being enhanced for future where it can detect our car arriving and automatically open the door and close once the car is safely parked. Researchers might as well think about adding the safety feature in this to avoid any car thefts. We have seen in offices where the light sensor switches on and off the lights by detecting the footsteps of people walking or any movement in the area. This helps to save energy and guides us through the dark places easily by lighting them up. There are smart homes in which all the devices are interconnected and can be controlled through the mobile phone from a very far off place; far can be as far as a different country.

Smart home can help all possible appliances to be connected using internet to form a Social Internet of Things network. We can have sensors and RFID in the switch boards, lights, kitchen appliances, entertainment corner—TV, music systems, smoke and gas leakage sensors, smart doors and windows with smart color changing glass or blinds, household appliances like air-conditioners, heaters, washing machines, etc. These can be connected to each other as objects and send data to our mobile phone apps for interactions. With a single click, we can access any of these devices (objects) from a remote location.

Different Issues in SIoT

In this section we place some of the issues present in SIoT:

- *Openness*: Beyond exhibiting crude information and other specific services, an IoT stage can likewise be adaptable enough to permit third gatherings to create complex applications through the procurement of an API.

- *Viability*: This property envelops two ideas—plan of action (whether it is suitable to market this engineering) and merchant lock-in (whether an organization can take the long haul danger of relying upon a specific supplier).

- *Reliability*: Not just the SIoT, structural planning must be strong enough to guarantee a certain level of accessibility, while providing an execution that is custom-made to the particular needs of the applications.

- *Scalability*: Within this standard, it is normal that the number of gadgets and the measure of information created prepared by these gadgets will develop exponentially (i.e., the idea of 'information deluge'). In this manner, we need to take adaptability and extensibility into record.

- *Interoperability*: Regardless of the possibility that the Social Internet of Things is innately heterogeneous, all its segments must be capable to communicate with one another. Subsequently, it is vital to attain administration and semantic interoperability, among different things.

- *Data Management*: As the distinctive components of the Social Internet of Things produce information, either by sensing or by transforming, we must take certain outline choices to store the data and the access mechanism.

- *Security Issues*: There are certain security issues that must be considered to accomplish a secure SIoT. In a secure SIoT, privacy of a patient should be maintained and data should be secure from local use.

One of the significant difficulties that must be overcome in request to push the Social Internet of Things into this present reality is security. IoT architectures should manage an evaluated populace of billions of items, which will connect with one another and with different elements, for example, people or virtual elements. Furthermore, all these communications must be secured some way or other, secure the data and administration provisioning of all important on-screen characters and limit the number of episodes that will influence the whole IoT. In this chapter we will try to pursue the existing security mechanism.

- Objects are heterogeneous in nature. Varied kinds of objects from hairpin to a bottle can be connected. In the same environment, there can be homogenous objects as well. This is where we think objects might conflict unless the naming of the objects is done properly (Nitti et al. 2015).

- Naming convention is numbering the objects to get a subjective as well as objective details of the objects. Redundancy needs to be checked upon to get a better trustworthiness of the object patterns.

- There are conditions where a person is located in India and is travelling to London. Local objects are connected where we do not need a WAN to interact. This is where the network interference might not be required. However, if objects are globally located, we might come across different networks which overlap. Trustworthiness by maintaining cloud data locally would keep the data untouched until it is triggered for a data.

- Tracking of data globally might become difficult; so we need to get local databases in cloud which can be easily maintained.

- If it is a small area, performances of these objects and data can be easily recorded. If globally data is stored, it might get infected.

- With the help of the proposed solution, redundancy can be avoided as we will consider maintaining the local nodes and objects and with the help of regular interval, S–special objects can be contacted for this information.

Trust in SIoT, Related Attacks and Trust Models

Using long-range interactive communicating objects in the SIoT to enable articles to self-sufficiently build up social connections is picking up the trend in recent years. The driving inspiration is that a social-situated approach is relied upon to put forward the revelation, determination and organization of administrations apart from data given by disseminated questions and systems that get to the physical world. Inside the subsequent protest interpersonal organization, a key object is to distribute data and administration, discover them and find novel assets to help the usage of complex administrations and applications. But, there has been one thing missing from many of the SIoT discussions to date is trust. This can be accomplished in a productive route by exploring 'friend' objects, rather than depending on the tools available over the Internet that can't scale to billions of future gadgets.

The SIoT represents an entirely different level of scale and complexity when it comes to the application of foundations for this trust, namely security and privacy. Every single device and sensor in the SIoT represents a potential risk. How confident can an organization be that each of these devices has the controls in place to preserve the confidentiality of the data collected and the integrity of the data sent? And one weak link could open up access to hundreds of thousands of devices on a network with potentially serious consequences. Considering the P2P network, the authors have provided a good analysis on the trustworthiness of the objects by providing a dynamic 'pre-trusted object' (DuBois et al. 2011).

In this section we mention the features of trust, trust-related attacks and classification of SIoT-trust models.

Trust in SIoT

Trust in its broadest form is having its existence in many spheres of life and in many branches of knowledge—from psychology to computer science. It is a multidimensional, multidisciplinary and multifaceted concept. It is a two-part activity involving a trustor and a trustee and the two parties are reliant on each other for their mutual benefit. They should be aware of the context of relationship, including purpose of trust and the environment comprising of time, location, activity, devices being used and the risk involved.

Some of the properties of trust as described in (Abdelghani et al. 2016) are as follows. Trust can be:

Direct: Trust is based on direct interactions, experiences or observations between the trustor and the trustee.

Indirect: The trustor and the trustee here don't have any past experiences or interactions. The trust here is built on the opinion and the recommendation of other nodes, which is called transitive trust (Josang 1997).

Local: It is limited to a pair of trustor and trustee. If one of them is changed or the roles are interchanged, the same trust may not hold.

Global: Over a period of time, a node might have developed a reputation in the form of level of trust in a community and it is irrespective of the other participating nodes (Troung et al. 2016, Xiao et al. 2015).

Asymmetric: In the normal sense of an asymmetric relation, the trust may or may not be both ways and even if it is both ways, the level may differ (Nitti et al. 2014).

Subjective: Trust is inherently a personal opinion, which is based on various factors or evidence, and that some of those may carry more weight than others (Grandison and Sloman 2000).

Objective: In some cases, it can be computed by using QoS properties of a device and some other factors using a certain formula.

Context-dependent: In a certain n a node i may have trust on another node j. But if the context changes, i may not trust j.

Composite property: Trust is really a composition of many different attributes: reliability, dependability, honesty, truthfulness, security, competence and timeliness, which may have to be considered depending on the environment in which trust has been specified (Grandison and Sloman 2000).

Dependent on history: The experience of a node in another similar context in the past may influence the present level of trust (Yan and Hollmann 2008).

Dynamic: The word trust has fuzziness, called informational fuzziness. So, it varies over periods of time. When more information reaches a node about the other node, then the first node may change its trust (Grandison and Sloman 2000).

Trust-related Attacks

Attacks may be carried out by malicious nodes in order to break the basic functionalities of IoT. These attacks are enumerated in (Abdelghani et al. 2016) as follows:

- *Self-promoting*: It can promote its importance (by providing good recommendations for itself), so as to be selected as the service provider, but then stops providing services or provides malfunction services.

- *Bad-mouthing*: It can ruin the reputation of well-behaved nodes (by providing bad recommendations against good nodes) so as to decrease the chance of good nodes being selected as service providers.

- *Ballot stuffing*: It can boost the reputation of bad nodes (by providing good recommendations for them) so as to increase the chance of bad nodes being selected as service providers.

- *Whitewashing*: A malicious node can disappear and re-join the application to wash away its bad reputation.

- *Discriminatory*: A malicious node can have discriminatory attack on non-friends or nodes without strong social ties (without many common friends) because of human nature or propensity towards friends in social IoT systems.

- *Opportunistic service*: A malicious node can provide good service to opportunistically gain high reputation, especially when it feels its reputation is dropping because of providing bad service. With good reputation, it can effectively collude with another bad node to perform bad-mouthing and ballot-stuffing attacks.

Classification of SIoT-trust Models

SIoT networks differ from the general social networks in features like; huge amount of nodes in the form of entities and devices involved, limited storage space, limited computation resources at these nodes, high dynamic structure due to frequent addition and deletion of nodes, high energy consumption, criticality and sensitiveness of user services.

A classification of SIoT was proposed as having four types of it as follows:

Trust Composition

There are two types of trust in the SIoT environment, called Quality of Service (QoS) which models the trust between a user and its devices and social trust which is trust between two users. Trust composition refers to the composition of factors in trust computation (Carminati et al. 2012).

QoS Trust: It depends upon competence, cooperativeness, reliability, task completion, capability, etc. of an IoT device in the form of service provided to a user.

Social Trust: It refers to the trust degree of the owners of IoT devices based upon intimacy, honesty, privacy, centrality and connectivity.

Trust Propagation

Trust can be passed on from nodes to other nodes, which we call trust propagation. It is categorised into:

Distributed propagation: No central entity is required for this type of propagation as IoT devices pass on their trust observations to another IoT device.

Centralised propagation: A central device takes care of propagation of trust observations.

Trust Aggregation

Aggregation is the process of collecting several components and generating a unique value out of these. Several approaches like static weighted sum, dynamic weighted sum, Bayesian model and Fuzzy Logic are used for the aggregation of trust in SIoT.

Trust Update

Trust values are dynamic as they change over time. These changes can be event-driven or time-driven. In event-driven updation, all trust data at a node is updated after a transaction or an event, whereas time-driven updation occurs periodically.

Current Status of Trust Management (Chen et al. 2011, Chen et al. 2015)

Honesty is the belief derived out of past experience of a node that another node is honest. A node is categorised as dishonest if it tries to put hurdles in the functionality and performance of the IoT network through improper recommendations. A dishonest node is treated as a malicious node. However, a node which does not cooperate in providing service as desired by another node is not counted as a malicious node. It can be considered as a selfish node, the ratio of the common interests of two nodes to the total number of group interests in the SIoT. Depending upon the three criteria of honesty, cooperative and community interest, Bao and Chen (Bao and Chen 2012) have incorporated social relationships into trust management for IoT. They framed direct trust management in SIoT as an aggregation of these three factors if the two nodes involved had interactions earlier. Otherwise, they depend upon indirect trust collected from other nodes which are related with them. This approach is too simple and does not take into consideration factors, like willingness of a node to cooperate and collaborate with other nodes. Also, the effectiveness of the proposed protocol on attacks against trust was not tested. The proposal did not take into consideration factors like storage management and energy consumption.

The same authors in Bao and Chen (2012) tried to improve their protocol by adding the features like scalability, adaptivity and survivability. A new storage management strategy is proposed in addition, which adds the scalability factor.

Another attempt (Bao et al. 2013) by the same set of authors proposed an adaptive and scalable trust management protocol for SOA-based IoT systems. The important point here is the addition of an adaptive filtering technique which is a function of direct and indirect trust to generate the overall trust, such that the convergence for trust evaluation is faster (Ashri et al. 2005). Here, the trust of the owners is taken into account instead of those for the devices. The trust value for all the devices and objects owned by specific persons are considered same, which does not take into account the QoS of the devices or objects.

Trustworthiness management in SIoT was dealt with in (Chen et al. 2015) through subjective and objective approaches. Both these approaches have some drawbacks in the form of slow transitive response and not being able to identify relation modifications of malicious persons. However, resiliency and scalability are enhanced by the use of distributed hash table.

The access service recommendation scheme for efficient service composition and malicious attacks assistance in SIoT environments was proposed in (Chen et al. 2015). It introduces an energy-aware mechanism for workload balancing and network stability. However, this approach suffers from having no specific solutions for scalability and limited storage and computational capacities of nodes.

A trust model is in which the three factors—reputation, recommendation and knowledge were incorporated by modulating human trust information process and social relationship. Reputation, which represents the global opinion of the trustee, recommendation, which depends upon suggestions from related nodes like friends, relatives, etc. and knowledge, which is based upon the information provided by the trustee through whom its trustworthiness can be assessed were used by the authors to create a trust model. This proposal does not cover the use of the protocol and the scalability of the SIoT.

In Christianson and Harbison (1997), a trust model is proposed which attaches every object with a reputation factor and can be updated by only the reputation server. Also, the objects are associated with their owner such that any addition or deletion of any object does not affect their baseline reputation. Positive and negative credits are allotted to the nodes for their correct and malicious services. Its drawback is that it only considers social relationships between human and attributes and attaches same trust value for all objects belonging to a specific owner.

The above protocols have one common drawback in the sense that they simply try to use the existing protocols to attain trust in traditional social networks. Because of the application for which these protocols were developed, the trust models for social networks have no provision for handling specific structures of SIoT and their scalability (Sethi et al. 2017). The relationships used are mostly human to human or object to object only.

Future Scope of Research

As has been discussed above, trust management in SIoT has many issues to handle, some of which are outlined in this section.

- Most of the work done so far has not focussed on trust management for human to object interaction. The other type of trust has been taken from IoT (for object to object interaction) or taken from social networks (human to human interaction). So, an extensive study on trust management for human to object interaction is necessary.

- Some kind of reward and penalty scheme has been developed, which makes the nodes alert. For malicious activities, the points awarded are negative. But, if a node judiciously shows the benevolent, then malicious responses can escape from low trust values. So, an in-depth study on the reward points is required.

- It has been assumed in evaluating trust that all the objects owned by an owner will be given the same trust values, irrespective of their entry point to the system. But this affects their relative trusts.

- Social relations management can be promoted, while collecting the trust values from each neighbouring node. So, study on the topic of social relationship promotion can be interesting.

Conclusions

Computers are still mostly dependent on human beings. As mentioned in Carminati et al. (2012), it is human beings, who feed data to the computers. Also, the human being is the sole creator of every single 'bit' of data available on the internet. Whatever data (right or wrong) is available for interaction between the 'things' has been initially fed by a human. Later, the tampering of data can be done as per requirement with the help of the computer. Human is still real and the data stored over the internet is still virtual. Objects do copy human's behaviour, virtually; however, RFID and sensors still cannot detect all the five senses of humans (Carminati et al. 2012a, Carminati et al. 2012b).

Trust and trust management play important roles in the functionality of SIoT. In this chapter, we started from scratch in defining concepts associated with trust management in SIoT, presented different attacks on trust, discussed issues involved in trust in SIOT, the current status of trust management and classification of trust models. Finally a few research directions in trust management of SIoT are presented.

References

Abdelghani, W., C. A. Zayani, I. Amous and F. Sedes. (2016). Trust management in social internet of things: A survey. Y. K. Dwivedi et al. (eds.) I3E LNCS 9844, pp. 430–441.

Ashri, R., S. Ramchurn, J. Sabater, M. Luck and N. Jennings. (2005). Trust evaluation through relationship analysis. pp. 1005–1011. *In*: Proc. 4th Int. Joint Conf. Autonomous Agents and Multiagent Systems. Utrecht, Netherlands.

Atzori, L., A. Iera and G. Morabito. (2011). SIoT: Giving a social structure to the internet of things. Communications Letters, IEEE 15(11): 1193–1195.

Atzori, L., A. Iera, G. Morabito and M. Nitti. (2012). The Social Internet of Things (SIoT); when social networks meet the internet of things: Concept, architecture and network characterization. Computer Networks 56(16): 3594–3608.

Bao, F. and I. R. Chen. (2012). Dynamic trust management for internet of things applications. pp. 1–6. *In*: Proceedings of the 2012 International Workshop on Self-aware Internet of Things, ACM.

Bao, F., I. R. Chen and J. Guo. (2013). Scalable, adaptive and survivable trust management for community of interest based internet of things systems. pp. 1–7. *In*: IEEE Eleventh International Symposium on De-centralized Systems (ISADS).

Carminati, B., E. Ferrari and J. Girardi. (2012). Trust and share: Trusted information sharing in online social networks. pp. 1281–1284. *In*: Proc. IEEE 28th ICDE, Washington, DC, USA.

Carminati, B., E. Ferrari and M. Viviani. (2012). A multi-dimensional and event-based model for trust computation in the social web. pp. 323–336. *In*: Social Informatics, Berlin, Germany: Springer.

Chen, D., G. Chang, D. Sun, J. Li, J. Jia and X. Wang. (2011). TRM-IoT: A trust management model based on fuzzy reputation for Internet of Things. Computer Sci. Inf. Syst. 8(4): 1207–1228.

Chen, R., J. Guo and F. Bao. (2015). Trust management for SOA-based IoT and its application to service composition. IEEE Transactions on Services Computing 2(1): 11.

Chen, Z., R. Ling, C. M. Huang and X. Zhu. (2015). A scheme of access service recommendation for the social internet of things. International Journal of Communication, Systems.

Christianson, B. and W. S. Harbison. (1997). Why isn't trust transitive? pp. 171–176. *In*: Proc. Int. Workshop Security Protocols, Cambridge, UK.

DuBois, T., J. Golbeck and A. Srinivasan. (2011). Predicting trust and distrust in social networks. pp. 418–424. *In*: Proc. IEEE 3rd Int. Conf. PASSAT and IEEE 3rd Int. Conference on Social Computing.

Fiske, A. P. (1992). The four elementary forms of sociality: Framework for a unified theory of social relations. Psychol. Rev. 99(4): 689–723.

Golbeck, J. A. (2005). Computing and Applying Trust in Web-based Social Networks. Ph.D. dissertation, Univ. Maryland, College Park, MD, USA.

Golbeck, J. and J. Hendler. (2006). Inferring binary trust relationships in web-based social networks. ACM Trans. Internet Technol. 6(4): 497–529.

Golbeck, J. (2008). Personalizing applications through integration of inferred trust values in semantic web-based social networks. p. 15. *In*: Proc. W8: Semantic Network Analysis.

Grandison, T. and M. Sloman. (2000). A survey of trust in internet applications. IEEE Communication on Survey Tutorials 3(4): 2–16.

https://www.linkedin.com/pulse/social-internet-things-future-smart-objects-michael-kamleitner.

https://blog.apnic.net/2015/10/20/5-challenges-of-the-internet-of-things/.

https://en.wikipedia.org/wiki/Amazon_Go.

http://searchhealthit.techtarget.com/essentialguide/A guide to healthcare IoT possibilities and obstacles.

Josang, A. (1997). Artificial reasoning with subjective logic. *In*: Proc. 2nd Australian Workshop Common Sense Reasoning, Perth, WA, Australia.

Kosmatos, E. A., N. D. Tselikas and A. C. Boucouvalas. (2011). Integrating RFIDs and smart objects into a unified Internet of things architecture. Adv. Internet Things 1(1): 5–12.

Li, H., B. Kyoungsoo and Y. Jaesoo. (2015). A mobile social network for efficient contents sharing and searches. Computers and Electrical Engineering 41: 288–300.

Liu, G., Y. Wang and L. Li. (2009). Trust management in three generations of web-based social networks. pp. 446–451. *In*: Proc. 2009 Symposium Workshops UIC-ATC, Brisbane, QLD, Australia.

Nitti, M., R. Girau, L. Atzori, A. Iera and G. Morabito. (2012). A subjective model for trustworthiness evaluation in the social Internet of things. pp. 18–23. *In*: Proceedings of the IEEE 23rd Int. Symposium PIMRC, Sydney, NSW, Australia.

Nitti, M., R. Girau and L. Atzori. (2014). Trustworthiness management in the social Internet of Things. IEEE Transaction on Knowledge and Data Engineering 26(5): 1253–1266.

Nitti, M. (2014). Managing the Internet of Things based on its Social Structure, Ph.D. Dissertation in Electronic and Computer Engineering Dept. of Electrical and Electronic Engineering University of Cagliari.

Nitti, M., L. Atzori and I. P. Cvijikj. (2015). Friendship selection in the social internet of things: challenges and possible strategies. IEEE Internet of Things Journal 2(3): 240–247.

Roman, R., P. Najera and J. Lopez. (2011). Securing the Internet of Things. Computer 44(9): 51–58.

Sethi, P. and S. R. Sarangi. (2017). Internet of Things: Architectures, protocols, and applications. Journal of Electrical and Computer Engineering.

Truong, N. B., T. W. Um and G. M. Lee. (2016). A reputation and knowledge-based trust service platform for trustworthy social internet of things. Innovations in Clouds, Internet and Networks (ICIN), Paris, France.

Xiao, H., N. Sidhu and B. Christianson. (2015). Guarantor and reputation based trust model for social internet of things. pp. 600–605. *In*: Wireless Communications and Mobile Computing Conference (IWCMC), 2015 International, IEEE.

Yan, Z. and S. Hollmann. (2008). Trust modelling and management from social trust to digital trust. IGI Global pp. 290–323.

Zhang, Y., J. Wen and F. Mo. (2014). The application of Internet of Things in social network. Computer Software and Applications Conference Workshops (COMPSACW), 2014 IEEE 38th International, IEEE.

CHAPTER **12**

Big Data Analysis
Technology and Applications

Abhijit Suprem

Introduction

Big data is a nebulous term that incorporates a variety of areas—from large-scale data collection, to storage methodologies, to analytics, to visualization. In each of these cases, the challenge is on efficient data operation on huge amounts of data. As computing power, hardware and storage continues to increase, there is no clear indication on what exactly the 'big' in big data means. Commonly, however, big data refers to giga-, tera-, or peta-scale data, such as text corpora from millions of books, the billions of images in Facebook's database, or the trillions of file signatures in security companies' malware identification databases.

Large-scale data collection is common in both industry and the public sector, and the presence of various collection agencies as financial entities, social media corporations, and public- and industry-sector monitoring organizations has significantly increased the volume of data collected and made publicly available. Modern research and development in big data goes beyond the collection and management paradigm and enters the domain of visual and data analytics. The former is concerned with effective interaction and visualization tools for developers and end-users to better analyze and tune tools, respectively. The latter deals with myriad

School of Computer Science, Georgia Institute of Technology, USA; asuprem@gatech.edu domain-specific frameworks, platforms and algorithms for a variety of analytics applications including data mining, prediction, ranking, language processing, financial modeling, human-computer interaction, and automated summarization. This survey covers the two broad research areas—visual analytics and data analytics, and details current trends and research in information visualization and large-scale analytics.

This paper provides a survey of big data tools and systems. We cover the data analytics pipeline, from data collection through crowdsourcing to data exploration and visualization. We then describe the two major parallelized data analytics models: MapReduce and Spark. The rest of the paper discusses various analytics applications for big data, such as recommender systems, graph analytics tools, cyber security, and social networks.

Crowdsourcing in Big Data

There is a significant amount of labeled data available in various domains, including financial transactions (Baviera et al. 2001), generalized object detection (Everingham et al. 2010), machine translation (Luong and Manning 2016), edge detection (Martin et al. 2001), etc. Unfortunately, there are many learning tasks for which the appropriate datasets do not exist and must be manually labeled. However, data labeling at scale requires enormous man-hours to complete. Crowdsourcing is one such approach that has been utilized to generate large datasets for machine and deep learning applications. We present some examples.

Visual Genome

The visual genome project presents a database of images with associated captions, region descriptions, region-bounding boxes, object labels and grounded scene graphs (Krishna et al. 2017). The dataset itself contains more than 108 K images. Manual labeling by the authors would be impossible; instead, the authors use Amazon's Mechanical Turk (Buhrmester et al. 2011) to crowdsource high-fidelity labels. For each image, human workers are tasked with multiple tasks sequentially to preserve fidelity. While (Krishna et al. 2017) goes into details on this process, we cover brief aspects. For node labeling, multiple humans are provided the same image and asked to label some objects within the image. Iterative applications where subsequent presentations of the image to other human workers also include prior labels, reduces the chance of the same labels being provided multiple times. Verification is also conducted by human workers, who are given worker-annotated labels and asked to verify whether the labels are correct. Multiple verifications for the same object are then combined to ensure that the label is indeed a correct label before being accepted. This process is repeated for each possible label type, i.e., bounding boxes, region captions, image captions, etc. to generate large-scale databases through distributed human labeling.

Visual Dialogue

The VQA dataset also uses Mechanical Turk in a similar fashion (Agrawal et al. 2017). Workers are provided images and asked to formulate questions about the image. Similar questions are discouraged by providing workers prior questions on the same image. In addition, each question goes through a validation stage before being accepted into the dataset. Answers are similarly generated by passing it through a dense labeling and verification stage as described.

Named Entity Recognition

This is a natural language processing task that traditionally is difficult to annotate as it may require expertise in a variety of fields to recognize named entities. A crowdsourced approach can lead to population sampling that can allow developers to collate annotations from multiple domain experts at a fraction of the cost. Lawson et al. (2010) proposes a named entity dataset for e-mails generated through such an approach. Each document is annotated by multiple users and high-quality annotations are incentivized to generate better annotations.

ReCaptcha

The ReCaptcha system is commonly used to provide gatekeeping across the Internet (No captcha recaptcha). It was designed to allow humans to solve it while stupefying machines by using human vision. This is an extension of crowdsourcing where an initially labeled set can be augmented. Each modern captcha consists of a question (How many of these are cars?), as well as a set of images—some with cars and some without. The images consist of novel and labeled images (unbeknownst to the user). The user annotations for the labeled images (i.e., ones ReCaptcha already knows are cars, in this case) are used as the basis of validating the users' other annotations; if the user correctly annotates the hidden known labeled set, then the annotations for the unlabeled sets are placed in consideration and the captcha is validated. These intermediate annotations from multiple users can then be collated to generate the correct labels.

Data Visualization

Data visualization, and by extension, visual analytics, are deeply intertwined with data analytics (Kohavi 2000). The purposes are two-fold—to develop useful large-scale visualizations that can view data-at-scale while still allowing users and developers to look at granular instances, and to allow top-level user interactions utilizing the visual space as a data or model exploration canvas. We cover some visual analytics frameworks for Deep Learning, Machine Learning, Data Exploration and Graph Analytics.

Deep Learning

Deep learning is a neural-network based technique for state-of-the-art results image/object recognition (Krizhevsky et al. 2012), speech synthesis (Wu and King 2016), machine translation (Luong et al. 2015), and a variety of applications far too numerous to cover here. We consider only challenges that necessitate data visualization. Deep learning models are, like most neural networks, black-boxes. It is, however, necessary to be able to diagnose or analyze deep learning tools to better understand model failure, training, efficacy and status, particularly for convolutional neural networks—a staple of image-based deep learning systems (Krizhevsky et al. 2012, Shvachko et al. 2010, Long et al. 2015). Tensorflow, a deep learning framework provided by Google (Abadi et al. 2016), contains the Tensorboard—a visual interface to examine model networks, performance and training status, each of which can be integral in developing deep learning models. CNNVis (Liu et al. 2017) also aid in network exploration and propagation, pathway visualization to better understand the activation pathway for specific data instances, and for data classes. Visualizing existing networks can aid in visual comparison to state-of-the-art network architectures (for example, ResNets linear depth (He et al. 2016) versus GoogLeNets Inception blocks (Szegedy et al. 2015) versus the standardized VGG convolutional block (Simard et al. 2003)) and help users in mixing network architectures for rapid model prototyping and deployment. Deep learning model performance relies similarly on visual examination. Model performance increases during training is not indicative of better models as convolutional neural networks are prone to overfitting (Smoot et al. 2010); as such, visual model performance on training and validation can better pinpoint crossover epochs where models begin to overfit for efficient model extraction. Finally, examining propagation pathways can help identify kernel redundancies in deep learning models while also pointing out concentrated kernels and low-activation irrelevant kernels for network pruning and subsequent compression.

Machine Learning

Machine learning visualization is a rich field incorporating myriad learning frameworks for data analytics, algorithm performance, and results exploration. We cover two tools in detail and refer the reader to (Simonyan and Zisserman 2014, Frank et al. 2009, Maaten and Hinton 2008) and (Donahue 2014) for details on other tools.

Podium: Podium (Wall et al. 2017) has been recently proposed as a ranking tool, for, e.g., SVM ranking. Podium allows users to specify a sample ranking for a subset S of data points out of the entire dataset N ($|S| << |N|$) and after generating rankings, allows for visual exploration of ranking weights and their contributory features from the dataset. A key contribution of Podium is to abstract high specific user-sourced ranking towards more generic holistic rankings. This allows users to use context and domain knowledge to provide the correct top-level rankings without tuning feature weights to tune the rankings.

Utopian: A tool for topic modeling, Utopian (Choo et al. 2012) provides semi-supervised inter-active user-driven approach that allows end-users to use domain knowledge and ne-tune the algorithms. The paper demonstrates the efficacy of incorporating user feedback directly into machine learning tools as front-end visual interfaces.

Facets: Introduced by Google, Facets (Wexler 2017) is a data exploration tool developed as a preliminary step to applying machine learning techniques. Key contributions include data cleaning and merging, as well as class-based visualizations and statistics within the dataset. Facets also includes 2D and quantile clustering visualizations to better help users understand their data. The former, as will be discussed below, can aid in identifying substructures within a dataset. The latter is a histogram-based representation that can identify key characteristics of data including representative centroids.

Data Exploration

Data exploration and pure visualization are one of the first steps in the data analytics pipeline. 2D clustering through dimensional reduction (Maaten and Hinton 2008) is a common technique for grouping similar data points in high dimensions. It can aid in preprocessing, outlier identification, and machine learning algorithm fine-tuning. Such dimensionality reduction techniques (i.e., Principal Component Analysis (Wold et al. 1987) and Linear Discriminant Analysis (Balakrishnama and Ganapathiraju 1998)) are generalized. Clustering visualization is a special case where high dimensional data is reduced to either two or three dimensions. Higher-dimension reductions, viz. reducing to greater than 3 dimensions allows for feature ranking to identify higher weight and lower weight features within the dataset (Wall et al. 2017). Graphical techniques include shape, hue and color encoding to represent multi-dimensional data, and interactive brushing for statistical exploration (Chen 2002). These techniques are commonly applied through front-end tools, such as tableau for data exploration and backend tools such as D3 for data visualization. D3 is a JavaScript library that uses dynamic data-bindings to produce interactive visualizations. It is a common tool in data visualization for a variety of visual techniques, from clustering (Maaten and Hinton 2008, Wold et al. 1987, Balakrishnama and Ganapathiraju 1998) to TableLens (Pirolli and Rao 1996, Pienta et al. 2014) to choropleths (Tobler 1973). Data, or information, visualization is applicable in all stages of the data analytics pipeline—from data exploration to cleaning to analysis to presentation.

Data Analytics

We cover the two primary computation models for large-scale parallelized data analytics—MapReduce and Spark. Each approach is a fault-tolerant, atomized, cluster-based approach to data operations at scale. We cover basic principles and advantages.

MapReduce

MapReduce (Dean and Ghemawat 2008) is a parallelization model for large-scale data processing. Introduced by Google in 2004, it has since been incorporated as a core component of the Apache Hadoop framework (Holmes 2012), which also uses a derivation of Google File System (Ghemawat et al. 2003) format from 2003 as its operating system. The Hadoop implementation is known as HDFS Hadoop Distributed File System (Sarwar et al. 2001). The MapReduce model performs parallel map and reduce operations on large datasets by splitting it into smaller clusters (commonly 64 MB) and passing them to generic computing workers. MapReduce incorporates fault tolerance through redundancies and handshakes multiple workers are sent the same to reduce long-tailed execution: once a worker completes the map or reduce task, the other workers on the same task are freed to operate on any outstanding idle tasks. MapReduce works as follows:

Idle Tasks: The data is split into clusters by the master node. Note that the HDFS, i.e., the system itself maintains the data across the networked infrastructure with copies (usually 3). As such, the master node can select the machine housing each data cluster to perform the tasks to reduce intra-network communications, and therefore, latencies. Each task is initialized as idle until assigned to a machine. Within limited machines, not all tasks may be assigned at first.

Map: The model assumes data is present as <key, value> pairs. The map operation collects all values for each key and passes the values, along with the key as an intermediate <key, list(values)> pair to the reduce (or the combine and reduce) operations.

Combine: In several domains and applications, the map step may produce significantly fragmented <key, list(values)> pairs. The canonical example is a word-frequency application that performs word counts across all documents. The map step can produce large numbers of the same <word, count> pair for each document (e.g., <the, 1>). The combine step performs some partial computations on each machine to preprocess the pairs for the reduce step. The reduce step takes the intermediate <key, list(values)> pairs and generates an output <key, value> pair for the specific application. In the provided canonical example, the reduce step may generate <the, 2454>, indicating 2454 instances of the word across all documents in the database. Figure 1 summarizes the MapReduce model.

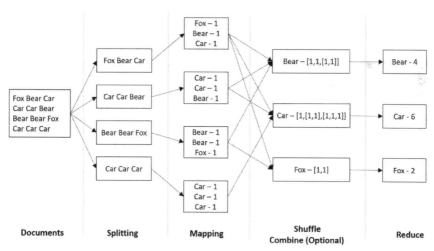

Fig. 1. The MapReduce model is described in steps. Documents: This is a collection of all documents in the file system. Splitting: The master node assigns each worker a document. Some documents may be idle. Mapping: The workers generate <key, value> pairs for each token. Shuffling: After each Map job is complete, all values for a key are collected. A Combiner step here may perform some of the summations. For example, the <key, value> for Bear with a combine step may be <bear, [1, 1, 2]> instead of <bear, [1, 1, [1, 1]]>. Reduce: Finally the master assigns reduce jobs, which sum the instances of words and reduce the word counts for each word.

Spark

Spark (Zaharia et al. 2010) is an alternative parallelization framework to MapReduce that operated on the Hadoop system (HDFS). A key difference is in-memory computations in Spark, which allow faster job completion at the expense of higher memory requirements. Where a MapReduce job can be split amongst various machines that each complete a portion of the map and reduce task, sequentially, a Spark job is completed in-memory with significant operation preprocessing. Spark allows for faster computation times by storing intermediate steps in memory instead of writing to disk as MapReduce does. Fault tolerance in Spark is incorporated by maintaining the set of data operations between intermediate data. The intermediate operations are a set of composed Scala functions that are compiled as a directed acyclic graph. As such, each intermediate step can be recreated from existing backups by following the cached steps for transformation.

Recommendation in Big Data

Recommender systems have become popular filtering methods for predicting user-rankings in a variety of applications, including Amazon product displays (Sanchez et al. 2009) and Netflix (Bennett et al. 2007) movie recommendations, and YouTube video recommendations (Davidson et al. 2010) to name a few. The motivation for recommender systems lies in allowing such companies to use prior user interactions, reviews and feedback to identify plausible future interactions and reviews for the users with other offerings; in the two examples given, Amazon may provide products similar to previously purchased or considered by the user, while Netflix may recommend media previously consumed by the user, while pushing irrelevant or disliked content further down per-user rankings. A particular technique is collaborative filtering (Sarwar et al. 2000), which relies on feature similarity to identify item rankings and provide recommendations.

Nearest Neighbor

Item-based nearest neighbor search (Cover and Hart 1967) and its variations (k-NN) (Peterson 2009, Keller et al. 1985) and approximate nearest neighbor (Indyk and Motwani 1998, Andoni and Indyk 2006) are useful approaches to collaborative filtering for finding objects with features similar to a query. These can be extended to multi-query nearest neighbor either by identifying query clusters by k-means clustering (Hartigan and Wong 1979) or other supervised or unsupervised clustering algorithms (Balcan et al. 2009) and finding nearest neighbors within these clusters, or taking learned weighted averages of various features to identify user centroids in the data space.

User-based nearest neighbor methods consider the user-space to apply nearest neighbors. Given a user u and a target item i, user-based neighborhood methods consider the set of all users U who recommend item i. Items where the user u is closer to the aggregate user space U_i of all users who recommend item i are placed higher in the recommendation rankings, while items whose user space U is dissimilar to the current user i are pushed lower. Unfortunately, most nearest neighbors approaches suffer heavily from the curse of dimensionality (Indyk and Motwani 1998, Friedman 1997) and are therefore infeasible for very high dimensional data. In addition, rating sparsity (i.e., users do not rate all movies) limits the feature selection for nearest neighbors.

Singular Value Decomposition

SVD-based approaches mitigate some of the problems with a nearest neighbors approach, namely high dimensionality and ratings sparsity (Cremonesi et al. 2010). Further, Stochastic Gradient Descent can lead to improved processing times. Appropriate regularizers can further regulate component matrices for sparsity and norm (Tang et al. 2013). SVDs key approach is to consider the rating of a particular user-movie pair as the dot product of a user vector u and a movie vector m. One particular approach we consider to provide a higher-level overview is PureSVD (Cremonesi et al. 2010). PureSVD considers a ratings matrix described as matrix product of a user matrix $\mathbf{U}$, a diagonal singular value matrix $\mathbf{\Sigma}$, and a movie matrix $\mathbf{Q}$. The pairwise prediction rule trivially follows, and a minimization is provided in (Anastasiu et al. 2016).

$$\underset{P,Q}{\arg\min} \frac{1}{2} \| \mathbf{R} - \mathbf{PQ}^T \|_F^2 + \frac{\beta}{2} (\| \mathbf{P} \|_F^2 + \| \mathbf{Q} \|_F^2)$$

where $\mathbf{P}$ and $\mathbf{Q}$ are each constrained to be orthogonal.

Graph Analytics

The emergence of modern graph databases has led to an increased focus on graph analytics, and by extension, graph visualizations. Existing tools for graph visualization include Gephi (Bastian et al. 2009), GraphViz (Ellson et al. 2001), Cytoscape (Smith et al. 2012), as well as more general-purpose visualization tools, as Tableau (Hanrahan 2003). In addition, graph databases themselves include primitive graph visualization tools. Neo4J (Developers 2012), a major graph database management system, includes a web front-end for graph visualization and subgraph retrieval and exploration. However, it is limited in scope (Pienta et al. 2017). The other tools mentioned, i.e., Gephi, GraphViz, and Cytoscape, include such operations as basic graph statistics, like graph density, diameter, average degree, to name a few; analysis, interactive layout tools such as node expansion, graph attributes, data binding for node radius, edge width, and edge length; and most important, graph display.

Visage

Recent research into graph visualization has also produced Visage (Pienta et al. 2017) a tool for graph visualization and pattern matching. Visage includes features as graph autocomplete an approach towards suggesting possible queries given partial graph queries by considering approximate matches and existing partial query results. In addition, Visage also includes support for user-sourced approximate matches using query wildcards by incorporating such prior work as G-Ray (Tong et al. 2007) and MAGE (Pienta et al. 2016). Such query wildcards extend current querying by allowing for wild-card based approximate matching. This is especially useful in clique search to find node associations. Graph autocomplete and wildcard matching are necessary for interactive query construction. In big data, this can be especially useful as users may not know the domain or range of data; graph. Autocomplete can suggest queries for data exploration while simultaneously narrowing the search space to plausible queries. In addition, query ranking can also play a useful role considering the partial query itself and node associations and proximity to the entire suggestion space. It is possible to rank autocomplete queries for suggestion. We can consider proximity by either node-to-node distance, or more probabilistically, a node proximity metric derived from random walk with restart (Tong et al. 2006) (note that this is also a core component of both G-Ray and MAGE, itself an extension of G-Ray).

Vigor

An extension of Visage is Vigor (Pienta et al. 2017) which operates as a tool for interactive graph querying and multi-coordinate data exploration and visualization. A key contribution of Vigor is a feature-independent subgraph that generates a 2D-clustering of high dimensional graph attributes. We suggest the canonical example from Vigor as a demonstrative example with conference publications. In addition, Vigor also incorporates Visages pattern matching with wildcards though approximate query expansion matching is not implemented.

Cyber Security

Cybersecurity in big data is a busy area. There are several research thrusts, including deep learning attacks and defense, fraud pattern detection, and malware identification. We cover representative examples of each.

Deep Learning Attacks

Attacks against deep learning models involve generating adversarial inputs that are designed to elicit attacker-sourced responses from the model (Papernot et al. 2016). Deep learning models rely on a complex interconnect of hidden layer neurons to generate classification outputs. The large parameter numbers make it possible to generate inputs programmatically that can force a convolutional neural network to recognize even artificial and unnatural images as a valid input (Papernot et al. 2016). In addition, Athalye and Sutskever (2017) show that such pixel-based manipulations, while invisible to the naked eye, can fool deep learning models into classifying the input into any variety of outputs. Deep learning models are traditional blackbox models whose inner workings are difficult to diagnose and analyze. As such, it is difficult to protect against such attacks as the avenues of attacks are difficult to identify. Visualization tools, like CNNVis (Liu et al. 2017) may help in finding compromised kernels for specified attacks. Output layer confidence intervals can also be examined to determine possibility of adversarial images (Feinman et al. 2017). In addition, Das et al. (2017) shows a simple technique for vaccinating deep learning models: image compression can reduce high frequency noise within the image and can lead to better classification results against adversarial inputs.

Fraud Pattern Detection

Fraud detection in large-scale graph databases is especially useful in such applications as financial markets, social networks and bank transactions. Supervised fraud detection can leverage such graph query tools as Neo4J, MongoDB and Titan to perform queries that can reveal fraudulent subgraphs, such as laundering rings or suspicious transaction activities (Castelltort and Laurent 2015). Large-scale graph analytics also prove useful, as per (Rao and Card 1995). The paper demonstrates the efficacy of identifying fraudulent networks through credit card association rules, which allow domain experts to identify suspicious versus good-faith associations between cards and users to better identify bad actors. In addition, Pandit et al. (2007) shows a similar approach using intuitive metrics of transaction frequencies and costs to identify reputable and non-reputable agents within an auction marketplace. The metrics are then converted to plausible features and node reputability is used to identify possible fraudulent transactions within the site.

Malware Detection

Traditional malware detection involves using program signatures and comparing them to a global malware signature database to identify possible cases of malicious software (Gri et al. 2009). However, recent improvements in attacks and hardware allow malicious actors to generate malware that may be different for each user (Bruschi et al. 2006), thereby providing a means to bypass the traditional malware detection. Techniques similar to fraud detection have been shown to work in such cases. Security companies maintain a global index of files on user systems. Each le is placed in an inverted index that identifies all machines that hold the file. While such an index would follow a heavy tail distribution, it is possible to use file associations to other files and machines to recognize malware. Tamersoy et al. (2014) provide an approach using such file-to-file associations, along with file features to identify malware. Chau et al. (2011) also show an association-based graph analytics technique to identify malware in tera-scale graphs.

Social Network Modeling

Large social networks are common with the ubiquitous use of platforms as Twitter, Facebook, YouTube, and, to a lesser extent, Google+. In addition, mobile Online Social Networks (Krishnamurthy and Wills 2010) are also prevalent, offering a variety of data from user-to-user interactions (Ellison et al. 2007), image and text sharing (through messages) and geolocation (Cranshaw et al. 2010).

Interaction Analysis

User interaction analysis is a common research thrust to identify useful interfaces that users can intuitively understand and apply. With the advent of data collection for user interactions-at-scale, it is possible to apply big data techniques towards online and offline user interaction analytics. The online model necessarily requires streaming models of computation and often scalable agnostic learning techniques that must make do with unlabeled or sketched data (Dasgupta et al. 2008). Agrawal et al. (2014) suggest that such large-scale interaction data can be processed to identify common user trends and behavioral patterns. Benevenuto et al. (2009) also present work in this area, focusing instead on click-level data. This is also an analogue of crowdsourced data collection, where user interaction is directly collected from end-users instead of user trials and focus groups. Such an approach mitigates sampling issues as the entire user base is necessarily representative of itself, and careful pruning of invalid data can maintain user demographics on training and test datasets. Benevenuto et al. (2009) also details interaction analytics. The paper analyzes online social network sessions and identifies key user activities within the silent and active interaction groups.

Network Modeling

Network modeling can also be incorporated to study information within social networks, as per (Adar and Adamic 2005), which studies meme propagation within blog networks. A meme, as coined in (Dawkins 2016), is a unit cultural element (such as a quote, text, image, or video) that is shared amongst users and can be modeled as self-replicating genes that carry cultural ideas. Adar and Adamic (2005) focus on subset links and studies its propagation across is blog subset. From a big data perspective, it becomes necessary to study such network modeling to better understand and act on information flow. There are considerable applications: categorizing modes of information flow to allow prediction of real-world events. Kallus (2014) and (Wang et al. 2012) suggest historical Twitter posts can be used to identify mass protests and civil disturbances as they occur, and may be used to predict mass protests based on information flow within and between user cliques. Jain et al. (2014) takes this further to demonstrate a viral event predictor by crawling Twitter activity and incorporating large-scale online graph analytics.

Approximate Search

Social networks lend trivially to graph-based representations (Greene and Cunningham 2013). As such, a graph query is a powerful method for retrieving user cliques and networks from a social network. While exact queries are possible, their implementation is readily available through such tools as Neo4J, MongoDB, and others. Approximate matching for a given query, on the other hand, is considerably more difficult (Tian and Patel 2008). Tong et al. (2007) demonstrate one possible approach to approximate matching, as well as an application in a corporate social network. Given a query, G-Ray uses a probabilistic proximity metric (Random Walk with Restar (Tong et al. 2006)) to perform approximate matching given a user query. MAGE (Pienta et al. 2016) is a more modern approach that allows for approximate matching with node and edge attributes, as well as wildcards, i.e., nodes and edges with any attributes;

the novel contribution being edge attributes included in the approximate match.

Social Media and Big Data

Social media analysis is the other side of the coin of social network modeling, where the latter aims to understand information flow between nodes in social network while the former aims to understand the information itself within the network. Such information can take many forms today, such as movie recommendations, product reviews, tweets, videos, photos and posts (Sing et al. 2005, Chen et al. 2012). As such, a wide variety of big data techniques apply in social media analysis.

Text Analytics

Text analytics is the process of retrieving structured predictions and conclusions from unstructured data. Text-based content, such as product reviews, recommendations and social media posts constitute large-scale corpora and thus fall under the big data domain. Approaches to text classification often incorporate recurrent neural networks (Fig. 2) unlike a convolutional neural network, which forms a directed acyclic graph.

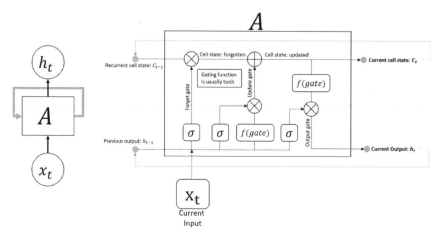

Fig. 2. A common RNN shows on the left, the network topology. x_t is the input at unit time t. h_t is the RNN output. A is the recurrent unit, described in detail on the right. Recurrent cell: has three major components: the forget gate, the update gate and the output gate. The forget gate is a multiplicative factor in [0, 1] applied to the previous cell state C_{t-1} to determine how much to 'forget'. A gate value of 0 indicates the current cell forgets everything. The update gate then applies a sigmoid and activation to the current input with the previous output and adds it to the current cell state. Finally, the output gate generates the current cell state for the next recurrence, as well as the current output.

Recurrent neural networks are cyclical and translate well to ordered domains, such as text and speech analytics and synthesis (Mikolov et al. 2010). We briefly cover major areas of text analytics.

Word Embeddings: A word embedding is a transformation from the vocabulary space to a high dimensional numerical vector in order to capture semantic relationships between words. While oracle lexical systems, such as WordNet (Miller 1995) can be used for such a purpose, their reliance on human subject matter experts reduces efficacy as they cannot be easily updated. Unsupervised word embeddings, however, are generated using word co-occurrences or n-grams intuitively. Words that have similar semantic meanings should appear in similar context (Turian et al. 2010). As such, normalized word co-occurrences provide a straightforward method for training word embeddings. Two widely used embeddings are word2vec (and its derivatives: sense2vec, lexvec, doc2vec, etc.) (Mikolov et al. 2013), developed by Google, and GloVe (Global Vectors for Word Embeddings) (Pennington et al. 2014). The difference between the two lies in the training process—word2vec is a predictive model relying on windowed n-grams to generate word vectors, while GloVe uses a factorized co-occurrence matrix on the entire corpora to generate its word vectors.

Part of Speech Tagging: An integral part of language processing and parsing is part-of-speech tagging to identify the roles each word or phrase plays in the document. Modern methods are stochastic; Wang et al. (2015) utilize bidirectional Long-Short Term Memory (biLSTM) re-current neural networks for this purpose. The BiLSTM-RNNs used by (Wang et al. 2015) can either be allowed to build their own word embeddings, or be provided existing word embeddings to speed up training and increase efficiency.

Sentiment Analysis: Sentiment analysis, or opinion mining, is a powerful method to gauge user perception and emotion (Pang et al. 2008). Steps include determining semantic orientation of phrases (either positive or negative), using oracle polarities of common words, such as bad or good, and using word embeddings or lexical databases to determine unknown word polarities by their proximity to known words. In addition, the presence of other numerical labels, such as reviews accompanied by a numerical score, such as a 5-star or 2-star review can also aid in determining word-polarity associations using an idea similar to co-occurrence-based word embedding training. We can observe that words that appear more often with lower review scores may indicate poor scores for future reviews (Turney 2002, Hatzivassiloglou and McKeown 1997). WordNet hierarchies and their analogues may also be used for such opinion detection and prediction (Esuli and Sebastiani 2005).

Geotagging

Image metadata in social media posts can provide detailed temporal footprints for users. Geospatial analytics use location metadata combined with other features, such as timestamp, posts, images, or sentiment analysis as described above to better understand relationships between location and user perception, i.e., whether certain sentiments of users indicate a willingness to be in a certain location, or whether time-of-day can be used to predict user location workplace. This lends to fine-grained geolocation using user-location history in conjunction with features mentioned to predict location, given a combination of such features, as per (Chong and Lim 2017).

Conclusion

There is little doubt that as processing powers, hardware capabilities, and storage quantities increase. The impact of big data will continue to be felt. In this paper, we have reviewed the data analytics pipeline, from crowdsourcing techniques, using Amazon Mechanical Turk, to Data Visualization tools for building, understanding, and ne-tuning analytics models and performing data exploration, to parallelized data analytics models, namely MapReduce and Spark. We also explored emerging graph analytics, and its applications in fraud detection, graph mining and approximate matching. With regard to applications, we cover various security aspects in the big data domain and possible preventive or vaccination measures against malicious actors. We also describe state-of-the-art approaches in convolutional neural network attacks, as well as graph-mining-based fraud and malware detection. Finally, we cover social network modeling and social media, and by extension, text analytics—already a major part of modern big data analytics. We describe modern text analytics approaches—from word embeddings to part-of-speech tagging to sentiment analysis. We note that there is significant emergent work in crowdsourced big data and neural networks as applied to pattern detection, speech and language synthesis, and object recognition. As better neural network architectures is discovered, and as better visualization tools for convolutional and recurrent neural networks (to name two) are developed, we expect application areas under such deep learning models to significantly increase.

Acknowledgments

We acknowledge Columbia University in the city of New York, where the first author obtained much of the knowledge and understanding of big data analytics, theoretical foundations of machine learning systems and data visualization through his MS degree at the Fu Foundation School of Engineering and Applied Sciences.

References

Abadi, M., A. Agarwal, P. Barham, E. Brevdo, Z. Chen, C. Citro, G. S. Corrado, A. Davis, J. Dean, M. Devin et al. (2016). Large-scale machine learning on heterogeneous distributed systems. arXiv preprint arXiv:1603.04467.

Adar, E. and L. A. Adamic. (2005). Tracking information epidemics in blogspace. pp. 207–214. *In*: Proceedings of the 2005 IEEE/WIC/ACM International Conference on Web Intelligence, IEEE Computer Society.

Agrawal, A., J. Lu, S. Antol, M. Mitchell, C. L. Zitnick, D. Parikh and D. Batra. (2017). Vqa: Visual question answering. International Journal of Computer Vision 123(1): 4–31.

Agrawal, D., C. Budak, A. El Abbadi, T. Georgiou and X. Yan. (2014). Big data in online social networks: User interaction analysis to model user behavior in social networks. In DNIS, Springer pp. 1–16.

Anastasiu, D. C., E. Christakopoulou, S. Smith, M. Sharma and G. Karypis. (2016). Big data and recommender systems.

Andoni, A. and P. Indyk. (2006). Near-optimal hashing algorithms for approximate nearest neighbor in high dimensions. pp. 459–468. *In*: Foundations of Computer Science, 2006. FOCS'06. 47th Annual IEEE Symposium on, IEEE.

Are you a robot? introducing "no captcha recaptcha", Dec 2014.

Athalye, A. and I. Sutskever. (2017). Synthesizing robust adversarial examples. arXiv preprint arXiv: 1707.07397.

Balakrishnama, S. and A. Ganapathiraju. (1998). Linear discriminant analysis-a brief tutorial. Institute for Signal and Information Processing, 18.

Balcan, M.-F., H. Roglin and S.-H. Teng. (2009). Agnostic clustering. pp. 384–398. *In*: ALT, Springer.

Bastian, M., S. Heymann, M. Jacomy et al. (2009). Gephi: an open source software for exploring and manipulating networks. ICWSM 8: 361–362.

Baviera, R., M. Pasquini, M. Serva, D. Vergni and A. Vulpiani. (2001). Correlations and multi-affinity in high frequency financial datasets. Physica A: Statistical Mechanics and its Applications 300(3): 551–557.

Benevenuto, F., T. Rodrigues, M. Cha and V. Almeida. (2009). Characterizing user behavior in online social networks. pp. 49–62. *In*: Proceedings of the 9th ACM SIGCOMM Conference on Internet Measurement Conference, ACM.

Bennett, J., S. Lanning et al. (2007). The netflix prize. pp. 35. *In*: Proceedings of KDD Cup and Workshop, New York, NY, USA.

Bruschi, D., L. Martignoni and M. Monga. (2006). Detecting self-mutating malware using control-ow graph matching. In DIMVA, Springer 6: 129–143.

Buhrmester, M., T. Kwang and S. D. Gosling. (2011). Amazon's mechanical turk: A new source of inexpensive, yet high-quality, data? Perspectives on Psychological Science 6(1): 3–5.

Castelltort, A. and A. Laurent. (2015). Fuzzy historical graph pattern matching a noSQL graph database approach for fraud ring resolution. pp. 151–167. *In*: IFIP International Conference on Artificial Intelligence Applications and Innovations, Springer.

Chau, D. H. P., C. Nachenberg, J. Wilhelm, A. Wright and C. Faloutsos. (2011). Polonium: Tera-scale graph mining and inference for malware detection. pp. 131–142. *In*: Proceedings of The 2011 Siam International Conference On Data Mining. SIAM.

Chen, C. (2002). Information visualization.

Chen, H., R. H. Chiang and V. C. Storey. (2012). Business intelligence and analytics: From big data to big impact. MIS Quarterly 36(4).

Chong, W. H. and E.-P. Lim. (2017). Exploiting contextual information for ne-grained tweet geolocation.

Choo, J., C. Lee, C. K. Reddy and H. Park. (2013). Utopian: User-driven topic modeling based on interactive nonnegative matrix factorization. IEEE Transactions on Visualization and Computer Graphics 19(12): 1992–2001.

Cover, T. and P. Hart. (1967). Nearest neighbor pattern classification. IEEE Transactions on Information Theory 13(1): 21–27.

Cranshaw, J., E. Toch, J. Hong, A. Kittur and N. Sadeh. (2010). Bridging the gap between physical location and online social networks. pp. 119–128. *In*: Proceedings of the 12th ACM International Conference on Ubiquitous Computing, ACM.

Cremonesi, P., Y. Koren and R. Turrin. (2010). Performance of recommender algorithms on top-n recommendation tasks. pp. 39–46. *In*: Proceedings of the Fourth ACM Conference on Recommender Systems, ACM.

Das, N., M. Shanbhogue, S.T. Chen, F. Hohman, L. Chen, M. E. Kounavis and D. H. Chau. (2017). Keeping the bad guys out: Protecting and vaccinating deep learning with jpeg compression. arXiv preprint arXiv: 1705.02900.

Dasgupta, S., D. J. Hsu and C. Monteleoni. (2008). A general agnostic active learning algorithm. pp. 353–360. *In*: Advances in Neural Information Processing Systems.

Davidson, J., B. Liebald, J. Liu, P. Nandy, T. Van Vleet, U. Gargi, S. Gupta, Y. He, M. Lambert, B. Livingston et al. (2010). The youtube video recommendation system. pp. 293–296. *In*: Proceedings of the Fourth ACM Conference on Recommender Systems, ACM.

Dawkins, R. (2016). The Selfish Gene. Oxford University Press.

Dean, J. and S. Ghemawat. (2008). Mapreduce: simplified data processing on large clusters. Communications of the ACM 51(1): 107–113.

Developers, N. (2012). Neo4j. Graph NoSQL Database [online].

Donahue, J., Y. Jia, O. Vinyals, J. Ho man, N. Zhang, E. Tzeng and T. Darrell. (2014). Decaf: A deep convolutional activation feature for generic visual recognition. pp. 647–655. *In*: International Conference on Machine Learning.

Ellison, N. B., C. Steinfield and C. Lampe. (2007). The benefits of facebook friends: social capital and college students use of online social network sites. Journal of Computer-Mediated Communication 12(4): 1143–1168.

Ellson, J., E. Gansner, L. Koutso os, S. C. North and G. Woodhull. (2001). Graphviz: open source graph drawing tools. pp. 483–484. *In*: International Symposium on Graph Drawing, Springer.

Esuli, A. and F. Sebastiani. (2005). Determining the semantic orientation of terms through gloss classification. pp. 617–624. *In*: Proceedings of the 14th ACM International Conference on Information and Knowledge Management, ACM.

Everingham, M., L. Van Gool, C. K. I. Williams, J. Winn and A. Zisserman. (June 2010). The pascal visual object classes (voc) challenge. International Journal of Computer Vision 88(2): 303–338.

Feinman, R., R. R. Curtin, S. Shintre and A. B. Gardner. (2017). Detecting adversarial samples from artifacts. arXiv preprint arXiv: 1703.00410.

Frank, E., M. Hall, G. Holmes, R. Kirkby, B. Pfahringer, I. H. Witten and L. Trigg. (2009). Weka-a machine learning workbench for data mining. pp. 1269–1277. *In*: Data mining and Knowledge Discovery Handbook, Springer.

Friedman, J. H. (1997). On bias, variance, 0/1 loss, and the curse-of-dimensionality. Data Mining and Knowledge Discovery 1(1): 55–77.

Ghemawat, S., H. Gobio and S.T. Leung. (2003). The google le system. pp. 29–43. *In*: ACM SIGOPS Operating Systems Review, vol. 37, ACM.

Greene, D. and P. Cunningham. (2013). Producing a unified graph representation from multiple social network views. pp. 118–121. *In*: Proceedings of the 5th Annual ACM Web Science Conference. ACM.

Gri, K., N. S. Schneider, X. Hu and T.-C. Chiueh. (2009). Automatic generation of string signatures for malware detection. pp. 101–120. *In*: RAID, vol. 5758, Springer.

Hanrahan, P. (2003). Tableau software white paper-visual thinking for business intelligence. Tableau Software, Seattle, WA.

Hartigan, J. A. and M. A. Wong. (1979). Algorithm as 136: A k-means clustering algorithm. Journal of the Royal Statistical Society. Series C (Applied Statistics) 28(1): 100–108.

Hatzivassiloglou, V. and K. R. McKeown. (1997). Predicting the semantic orientation of adjectives. In Proceedings of the Eighth Conference on European Chapter of the Association for Computational Linguistics. pp. 174–181. Association for Computational Linguistics.

He, K., X. Zhang, S. Ren and J. Sun. (2016). Deep residual learning for image recognition. pp. 770–778. *In*: Proceedings of the IEEE Conference on Computer Vision and Pattern Recognition.

Holmes, A. (2012). Hadoop in practice. Manning Publications Co.

Indyk, P. and R. Motwani. (1998). Approximate nearest neighbors: towards removing the curse of dimensionality. pp. 604–613. *In*: Proceedings of the Thirtieth Annual ACM Symposium on Theory of Computing, ACM.

Jain, P., J. Manweiler, A. Acharya and R. R. Choudhury. (2014). Scalable social analytics for live viral event prediction. In ICWSM.

Kallus, N. (2014). Predicting crowd behavior with big public data. pp. 625–630. *In*: Proceedings of the 23rd International Conference on World Wide Web, ACM.

Keller, J. M., M. R. Gray and J. A. Givens. (1985). A fuzzy k-nearest neighbor algorithm. IEEE Transactions on Systems, Man, and Cybernetics (4): 580–585.

Kohavi, R. (2000). Data mining and visualization. National Academy of Engineering (NAE), pp. 1–8.

Krishna, R., Y. Zhu, O. Groth, J. Johnson, K. Hata, J. Kravitz, S. Chen, Y. Kalantidis, L.-J. Li, D. A. Shamma et al. (2017). Visual genome: Connecting language and vision using crowdsourced dense image annotations. International Journal of Computer Vision 123(1): 32–73.

Krishnamurthy, B. and C. E. Wills. (2010). Privacy leakage in mobile online social networks. pp. 4–4. *In*: Proceedings of the 3rd Conference on Online Social Networks, USENIX Association.

Krizhevsky, A., I. Sutskever and G. E. Hinton. (2012). Imagenet classification with deep convolutional neural networks. pp. 1097–1105. *In*: Advances in Neural Information Processing Systems.

Lawson, N., K. Eustice, M. Perkowitz and M. Yetisgen-Yildiz. (2010). Annotating large email datasets for named entity recognition with mechanical turk. *In*: Proceedings of the NAACL HLT 2010 Workshop on Creating Speech and Language Data with Amazon's Mechanical Turk. pp. 71–79. Association for Computational Linguistics.

Liu, M., J. Shi, Z. Li, C. Li, J. Zhu and S. Liu. (2017). Towards better analysis of deep convolutional neural networks. IEEE Transactions on Visualization and Computer Graphics 23(1): 91–100.

Long, J., E. Shelhamer and T. Darrell. (2015). Fully convolutional networks for semantic segmentation. pp. 3431–3440. *In*: Proceedings of the IEEE Conference on Computer Vision and Pattern Recognition .

Luong, M.-T., H. Pham and C. D. Manning. (2015). Effective approaches to attention-based neural machine translation. arXiv preprint arXiv: 1508.04025.

Luong, M.-T. and C. D. Manning. (August 2016). Achieving open vocabulary neural machine translation with hybrid word-character models. In Association for Computational Linguistics (ACL), Berlin, Germany.

Maaten, L. V. D. and G. Hinton. (2008). Visualizing data using t-sne. Journal of Machine Learning Research 9: 2579–2605.

Martin, D., C. Fowlkes, D. Tal and J. Malik. (July 2001). A database of human segmented natural images and its application to evaluating segmentation algorithms and measuring eco-logical statistics. In Proc. 8th Int'l Conf. Computer Vision 2: 416–423.

Mikolov, T., M. Kara at, L. Burget, J. Cernocky and S. Khudanpur. (2010). Recurrent neural network based language model. In Interspeech 2(3).

Mikolov, T., I. Sutskever, K. Chen, G. S. Corrado and J. Dean. (2013). Distributed representations of words and phrases and their compositionality. pp. 3111–3119. *In*: Advances in Neural Information Processing Systems.

Miller, G. A. (1995). Wordnet: a lexical database for english. Communications of the ACM 38(11): 39–41.

Pandit, S., D. H. Chau, S. Wang and C. Faloutsos. (2007). Netprobe: a fast and scalable system for fraud detection in online auction networks. pp. 201–210. In Proceedings of the 16th International Conference on World Wide Web, ACM.

Pang, B., L. Lee et al. (2008). Opinion mining and sentiment analysis. Foundations and Trends R in Information Retrieval 2(1-2): 1–135.

Papernot, N., P. McDaniel, I. Goodfellow, S. Jha, Z. B. Celik and A. Swami. (2016). Practical black-box attacks against deep learning systems using adversarial examples. arXiv preprint arXiv: 1602.02697.

Papernot, N., P. McDaniel, S. Jha, M. Fredrikson, Z. B. Celik and A. Swami. (2016). The limitations of deep learning in adversarial settings. pp. 372–387. *In*: Security and Privacy (EuroS&P), 2016 IEEE European Symposium on, IEEE.

Pennington, J., R. Socher and C. Manning. (2014). Glove: Global vectors for word representation. pp. 1532–1543. *In*: Proceedings of the 2014 Conference on Empirical Methods in Natural Language Processing (EMNLP).

Peterson, L. E. (2009). K-nearest neighbor. Scholarpedia 4(2): 1883.

Pienta, R., A. Tamersoy, H. Tong and D. H. Chau. (2014). Mage: Matching approximate patterns in richly-attributed graphs. pp. 585–590. *In*: Big Data (Big Data), 2014 IEEE International Conference on. IEEE.

Pienta, R., A. Tamersoy, A. Endert, S. Navathe, H. Tong and D. H. Chau. (2016). Visage: Interactive visual graph querying. pp. 272–279. *In*: Proceedings of the International Working Conference on Advanced Visual Interfaces, ACM.

Pienta, R., F. Hohman, A. Endert, A. Tamersoy, K. Roundy, C. Gates, S. Navathe and D. H. Chau. (2017). Vigor: Interactive visual exploration of graph query results. IEEE Transactions on Visualization and Computer Graphics.

Pirolli, P. and R. Rao. (1996). Table lens as a tool for making sense of data. pp. 67–80. *In*: Proceedings of the Workshop on Advanced Visual Interfaces, ACM.

Rao, R. and S. K. Card. (1995). Exploring large tables with the table lens. pp. 403–404. *In*: Conference Companion on Human Factors in Computing Systems, ACM.

Sanchez, D., M. Vila, L. Cerda, and J.-M. Serrano. (2009). Association rules applied to credit card fraud detection. Expert Systems with Applications 36(2): 3630–3640.

Sarwar, B., G. Karypis, J. Konstan and J. Riedl. (2000). Application of dimensionality reduction in recommender system—a case study. Technical report, Minnesota Univ. Minneapolis Dept. of Computer Science.

Sarwar, B., G. Karypis, J. Konstan and J. Riedl. (2001). Item-based collaborative filtering recommendation algorithms. pp. 285–295. *In*: Proceedings of the 10th International Conference on World Wide Web, ACM.

Shvachko, K., H. Kuang, S. Radia and R. Chansler. (2010). The hadoop distributed le system. pp. 1–10. *In*: Mass Storage Systems and Technologies (MSST), 2010 IEEE 26th Symposium on, IEEE.

Simard, P. Y., D. Steinkraus, J. C. Platt et al. (2003). Best practices for convolutional neural networks applied to visual document analysis. In ICDAR 3: 958–962.

Simonyan, K. and A. Zisserman. (2014). Very deep convolutional networks for large-scale image recognition. arXiv preprint arXiv: 1409.1556.

Sing, T., O. Sander, N. Beerenwinkel and T. Lengauer. (2005). ROCR: visualizing classifier performance in R. Bioinformatics 21(20): 3940–3941.

Smith, M., C. Szongott, B. Henne and G. Von Voigt. (2012). Big data privacy issues in public social media. In Digital Ecosystems Technologies (DEST), 2012 6th IEEE International Conference on, IEEE pp. 1–6.

Smoot, M. E., K. Ono, J. Ruscheinski, P.-L. Wang and T. Ideker. (2010). Cytoscape 2.8: new features for data integration and network visualization. Bioinformatics 27(3): 431–432.

Srivastava, N., G. E. Hinton, A. Krizhevsky, I. Sutskever and R. Salakhutdinov. (2014). Dropout: a simple way to prevent neural networks from overfitting. Journal of Machine Learning Research 15(1): 1929–1958.

Szegedy, C., W. Liu, Y. Jia, P. Sermanet, S. Reed, D. Anguelov, D. Erhan, V. Van-houcke and A. Rabinovich. (2015). Going deeper with convolutions. pp. 1–9. In: Proceedings of the IEEE Conference on Computer Vision and Pattern Recognition.

Tamersoy, A., K. Roundy and D. H. Chau. (2014). Guilt by association: large scale malware detection by mining le-relation graphs. pp. 1524–1533. In: Proceedings of the 20th ACM SIGKDD International Conference on Knowledge Discovery and Data Mining, ACM.

Tang, Y., Y. Shen, A. Jiang, N. Xu and C. Zhu. (2013). Image denoising via graph regularized k-svd. pp. 2820–2823. In: Circuits and Systems (ISCAS), 2013 IEEE International Symposium on, IEEE.

Tian, Y. and J. M. Patel. (2008). Tale: A tool for approximate large graph matching. pp. 963–972. In: Data Engineering, 2008. ICDE 2008. IEEE 24th International Conference on, IEEE.

Tobler, W. R. (1973). Choropleth maps without class intervals? Geographical Analysis 5(3): 262–265.

Tong, H., C. Faloutsos and J.-Y. Pan. (2006). Fast random walk with restart and its applications.

Tong, H., C. Faloutsos, B. Gallagher and T. Eliassi-Rad. (2007). Fast best-effort pattern matching in large attributed graphs. pp. 737–746. In: Proceedings of the 13th ACM SIGKDD International Conference on Knowledge Discovery and Data Mining, ACM.

Turian, J., L. Ratinov and Y. Bengio. (2010). Word representations: a simple and general method for semi-supervised learning. pp. 384–394. In: Proceedings of the 48th Annual Meeting of the Association for Computational Linguistics, Association for Computational Linguistics.

Turney, P. D. (2002). Thumbs up or thumbs down?: semantic orientation applied to unsupervised classification of reviews. pp. 417–424. In: Proceedings of the 40th Annual Meeting on Association for Computational linguistics. Association for Computational Linguistics.

Wall, E., S. Das, R. Chawla, B. Kalidindi, E. T. Brown and A. Endert. (2017). Podium: Ranking data using mixed-initiative visual analytics. IEEE Transactions on Visualization and Computer Graphics.

Wang, P., Y. Qian, F. K. Soong, L. He and H. Zhao. (2015). Part-of-speech tagging with bidirectional long short-term memory recurrent neural network. arXiv preprint arXiv: 1510.06168.

Wang, X., M. S. Gerber and D. E. Brown. (2012). Automatic crime prediction using events extracted from twitter posts. SBP 12: 231–238.

Wexler, J. (2017). Facets: An open source visualization tool for machine learning training data.

Wold, S., K. Esbensen and P. Geladi. (1987). Principal component analysis. Chemometrics and Intelligent Laboratory Systems 2(1-3): 37–52.

Wu, Z. and S. King. (2016). Investigating gated recurrent networks for speech synthesis. pp. 5140–5144. In: Acoustics, Speech and Signal Processing (ICASSP), 2016 IEEE International Conference on, IEEE.

Zaharia, M., M. Chowdhury, M. J. Franklin, S. Shenker and I. Stoica. (2010). Spark: Cluster computing with working sets. HotCloud 10(10-10): 95.

CHAPTER **13**

Social Networking in Higher Education
Saudi Arabia Perspective

Waleed Khalid Almufaraj and *Tomayess Issa**

Introduction

It has been claimed by several research studies (Deng and Yuen 2009, Sturgeon and Walker 2009, Zakaria et al. 2013) that social networks (SNs) are engaging student interaction with the knowledge and learning in an informal way. Students are learning and acquiring more knowledge through SNs because they find the new informal approach more interesting than the traditional and formal classroom environment. Therefore, educational institutions are adopting SNs as an informal way of delivering knowledge and learning (Deng and Yuen 2009). Several research studies have claimed that SNs are influencing the ownership, experience, skills, and preferences of the students (Robbins-Bell 2008). Due to digital communication and interaction, students acquire and share knowledge in an informal way. Hence, students learn more in an informal way compared to the formal and traditional manner of receiving education (Schroeder and Greenbowe 2009). Due to such opportunities, several educational institutions have started to use Facebook as well as other social networks in order to train and teach students informally. Students use Facebook, Twitter,

Curtin University, School of Management, Perth Australia.
Email: w.almufaraj@graduate.curtin.edu.au
* Corresponding author: Tomayess.Issa@cbs.curtin.edu.au

LinkedIn and Myspace to interact with other students and engage in academic discussions (Selwyn 2009). This enhances the critical thinking and knowledge of the students. Moreover, a personalised relationship develops that motivates and encourages students to learn, argue, interact and have a greater involvement in educational activities (Sturgeon and Walker 2009).

Educational institutions implement SNs in various ways in order to enhance the learning experience of their students. Traditionally, SNs have been used to engage students in discussion forums, activity forums and social forums (Wankel 2009). These days, students are encouraged to use Facebook, Twitter, LinkedIn and other forms of SN. These modern forms of SNs are more effective for informal education because students perceive them as networks of fun rather than formal networks of academic discussion. As the perception of students about these SNs changes, their learning and knowledge increases. Therefore, several researchers have claimed that the modern informal SNs, like Facebook, are more effective in enhancing the learning and knowledge of students compared with the traditional rigid and academic forms of SNs, such as Wikis and discussion forums (Selwyn 2009). In Saudi Arabia, health and medical institutions have been the first to adopt SNs to provide informal medical education to students (Alanzi et al. 2014). By following the medical education sector, other higher education sectors are also acknowledging the significance of SNs and are adopting SNs for education and dissemination of knowledge (Zakaria et al. 2013). Certain forms of SNs are already available in Saudi Arabia as Saudi Arabian universities are using educational Wikis as SNs to enhance interaction among students within a university and between universities. This has also raised several challenges for educational institutions due to privacy and confidentiality of the information that is available in social networks and security of information systems. Higher education institutions are also faced with different challenges that may interfere with their information systems because of using social networking in their learning institutions (Al-Khalifa and Garcia 2013). In this research paper, the impact of social networks on students in higher education sector of Saudi Arabia is researched and analysed. The synopsis of the paper is as under: Introduction, social networks and higher education sector, the use of social networking in Saudi Arabian higher education sector, research gap, research questions and methods, participants, results, discussion and new significance, research limitations and conclusion.

Social Networks and Higher Education Sector

Veletsianos et al. (2013) argue that SN has had an impact on education and learning organisations as it provides an innovative means of engaging students in learning. They argue that universities have recognised SN as a powerful tool to communicate with students because it enables them to engage the students at a personal level while providing education. Therefore, the majority of universities in developing nations have introduced SNs in their campuses to enhance student knowledge and learning as well as to enhance communication and interaction with them.

Huang et al. (2013) state that educational use of SN has attracted the attention of researchers because SN has become a strong policy making tool to enhance student knowledge and learning experience. They claim that 50 per cent of students in the United States universities use on-campus social networks to interact and communicate with other students. Ellison et al. (2007) claim 73 per cent of the students in Europe are using social networks in their educational institutions to engage with their peers. They also state that universities are significantly adopting SN because it enables them to enhance professional education, curriculum education, and learning.

Sancho and Vries (2013) discussed the difference of opinion between developed and developing nations regarding the implementation of social networks in educational institutions. US researchers argue that social networks in higher education have enhanced the learning and academic knowledge of the students. On the other hand, researchers from Asia and Africa argue that social networks have negative effects on the academic life of students as their attention gets diverted from studies and they start actively engaging in social interactions rather than education. Hence, social networks act as a distraction for students from their learning. Lewis and Rush (2013) want SN-enabled universities to connect their students with professionals and professional networks. Several universities have allowed their students to create and use their LinkedIn and Twitter accounts on the campuses to connect with entrepreneurs, professionals and experts from the industry. As a result, when students interact and engage with these professionals, they obtain a practical orientation and application of their knowledge. In this way, SN is being used by universities as a professional development tool while ties with academic and professional lives is maximising the knowledge capability of the students.

McGuinnes and Marchand (2014) write that most of the universities use SNs to connect their current students with the alumni in the professional world. When students and peers share their interest, their knowledge and ideas are enhanced significantly and they get practical exposure to their peers and seniors in the professional field. As a result, a social capital is generated for the universities while students' academic performance is enhanced with personal relationship and exchange of information between students and professionals. In this way, SN is being used as an important tool to enhance the professional exposure of students. Dabbagh and

Kitsantas (2013) state that several universities are adopting SNs because they believe that SNs enhance the transferable, technical and social skills of the students due to informal and formal learning experience. They claim that students are exposed to formal learning environments from the start of their education, and by the time students reach the tertiary level, they are bored and disengaged.

Therefore, SNs provide informal learning experiences for students besides providing breakthroughs in enhancing their knowledge and capability. In a formal education environment, a student normally learns, but SNs encourage students to engage. With instant messaging, e-mailing and blogging, students are involved in the learning and interaction process that encourages their participation. In this way, students who usually would not participate within a formal learning environment can be engaged in class discussions and activities through a more informal learning approach. Students, therefore, can benefit from increased involvement and just-in-time learning.

Mitra and Padman (2014) state that SNs enable the tutors to deliver the prescribed curriculum more effectively than normal teaching methods. Students have a blend of fun and education that makes education and learning interesting for them. When they feel that they can obtain guidance through websites and from their homes, they treat learning as fun and they can access it at any time. As a result, because students' capabilities and motivation are enhanced, universities have been encouraged use of SNs to improve student engagement, participation and media skills. Gray et al. (2013) argue that SNs develop a participatory culture in educational institutions where students are continuously engaged and mentored through sharing, engagement and social interaction. As a result, informal knowledge of the students is enhanced because teachers can gain more insight regarding the former by engaging with them personally. Moreover, social networks allow students to collaborate and converse with each other at the same time. As a result, SNs have become a virtual space for learners. This virtual space acts as a gift economy for students who share and gift knowledge and learning to each other. Moreover, this knowledge is also gifted by the professionals and teachers when they interact with the students. Thus enhancement of informal learning increases students' knowledge and this has encouraged universities to integrate SNs in their pedagogy.

Erkollar and Oberer (2013) criticised the use of social networking in educational institutions and tertiary education. They argued that no educational benefit is offered by social networking; instead, students are engaged in wasting their time, breaching the privacy of others and bullying instead of educational learning. For example, they tested and analysed the impact of Google+ on education and knowledge of students. It was found that students used Google+ more for irrelevant, time-wasting activities rather than using it for education and learning. Moreover, most of the students were engaged in cyberbullying rather than adding some real value to class participation. Therefore, they state the access to social networking should be completely restricted in higher education universities and it is a mistake that universities are sanctioning and encouraging social networking on their campuses. Stack (2013) writes that several universities that used social networks on their campuses have decided to remove them because of cyberbullying and privacy theft issues. Instead of engaging in educational activities, students start engaging in cyber stalking, sexual harassment and cyber bullying. Hence, more harm than good is done. The incidence of depression among students increases because of loss of self-esteem and stress. Also, most of the discussions on social networking end up in disputes and contradictions that impact on the behaviour of the students. Therefore, Stack (2013) suggests that it is important to ban SNs instead of implementing them as an educational tool.

Ma and Au (2014) write that both advantages and disadvantages of social networking are being identified through academic discussion and analyses. It should be noted that most of the research studies focussed on the positive impact of SNs on students, and that this encouraged most universities to adopt social networking and implementing SNs rapidly. However, after implementation, the negative aspects of SNs soon emerged. This has led to a difference of opinion among academics about whether the advantages of social networking outweigh its disadvantages or vice versa. Despite the advantages and disadvantages of social networking in higher education, it is evident that its role in educational institutions is increasing significantly. This report will investigate the role of social networking in education and its advantages and disadvantages. Table 1 discusses and highlights the use of social network in educational institutions.

Use of Social Networking in Saudi Arabian Higher Education Sector

It should be noted that social networking in the higher education institutions of Saudi Arabia is significantly limited and there is a significant research gap in this context as well. The use of internet has been increasing since 1998 but it is primarily used for distance-learning programmes, particularly in medical schools. The healthcare sector of Saudi Arabia can be regarded as the first beneficiary of SNs in higher education and medical education sectors, where students significantly use social networks for sharing and gathering information and to interact with other students in order to seek information. Moreover, research studies claim that the Saudi

Table 1. Statistics and insights related to social network use in higher education.

Country	Social Network	Statistics	Insight
United States (U.S.A.) and Australia	Facebook and Twitter	Ninety-six per cent of the students in US and Australia spend 30 minutes daily on social networks such as Facebook and Twitter. Also, 80 per cent of the educators have a social network account as well (Chauhan and Pillai 2013).	This highlights the significance and importance of social networks that could be used to connect 96 per cent of the students with the educators.
U.S.A.	All Networks	Students between the ages of 18–25 spend one hour daily on social networking websites (Chauhan and Pillai 2013).	If social networks are adopted in higher education, then students are likely to spend this one hour on informal education rather than wasting their time (Cheung et al. 2011).
US and Europe	Wikis	Ninety-eight per cent of the students are more actively involved in social networks compared with participation in local college Wikis and networks (Gibb et al. 2013).	If educational institutions started using these social networks as a replacement for these Wikis, then a high response rate can be obtained from the students.
US	LinkedIn	LinkedIn has emerged as one of the largest professional networks to connect industry professionals. 40 per cent of students have a LinkedIn account as well which they use to connect with their peers and seniors working in industry (Das et al. 2012).	After analysing this opportunity, many universities have connected their social networks with their alumni and industry professionals to increase formal education of students (Greenhow 2011).

higher education sector benefits from social networking just the same as in other parts of the world (Alanzi et al. 2014).

Students in the larger universities of Saudi Arabia are benefitting significantly from these universities' increasing implementation of social networks. King Khalid University and King Saud University have introduced social Wikis to engage the students in informal collaborative learning. Similarly, these social networks are connected to the professional world that provides rapid access to students for professional information and knowledge. Students mostly use Twitter and YouTube in these universities in order to access professional information in an informal way. Moreover, there is an increasing awareness of the benefits and advantages of SNs in higher education in Saudi Arabia (Zakaria et al. 2013). Wiki is one of the most important and widely-implemented forms of social networks in Saudi Arabian universities. On the other hand, the use of other social networks like Facebook, YouTube and Twitter is limited in educational institutions. This is because the higher education sector in Saudi Arabia is still facing some resistance although it has recognised the significance and importance of social networks in higher education. Most of the educational institutions are confused due to the perceived risks associated with social networks that could potentially harm the students. The risks include issues associated with privacy and confidentiality of the information and the security of information systems (Al-Khalifa and Garcia 2013).

The political, economic, social and technological environment of Saudi Arabia is an important obstacle to the adoption of social networks in Saudi Arabia, which is an Islamic country and, therefore, in order to protect Islamic values that are their right, there are significant restrictions on information, internet, and online content because adult content is strongly prohibited in Islam and Saudi authorities never provide access to such content. This is their right but it is one of the obstacles to the current adoption of social networks as students have limited access to information in such a controlled environment. Secondly, the Saudi Arabian government, in accordance with Islamic teachings, discourages free interaction between men and women. Therefore, social networks like Facebook are heavily restricted in order to prohibit this interaction, and information access by students is closely monitored. These also restrict the freedom of expression in Saudi Arabia due to limited online access (Zakaria et al. 2013).

The main purpose of social networking in higher education is to provide an informal means of interaction and education to students, so that the latter gain and share information that can enhance their knowledge, understanding and professional exposure. Citizens' freedom of expression is limited in Saudi Arabia and social networks are closely monitored. Therefore, the laws that restrict this freedom should be relaxed a little so that students can benefit fully from social networks. This free access is important because students in higher education conduct different research that may be contrary to Islamic law but could help students to understand various concepts pertaining to human psychology and sociology (Alanzi et al. 2014).

This implies that students in higher education should be granted full access to social network content with permission from the communication and information technology commission to enable them to carry out their research. In this way, social networks will improve the learning outcomes in higher education and benefit society overall (Al-Khalifa and Garcia 2013). This researcher reviewed and researched various journals in an attempt to find studies related to social networks in higher education in Saudi Arabia; however, this information is not available, which clearly indicates a research gap. The opportunities offered by social networks in higher education as reported in different research studies are listed below:

1. Cutting edge knowledge (Dabbagh and Kitsantas 2013, Sturgeon and Walker 2009, Wankel 2009, Li and Pitts 2009).
2. Collaboration (Selwyn 2009, Schroeder and Greenbowe 2009, Deng and Yuen 2009).
3. Acquire new acquaintances (Boon and Sinclair 2009, Ras and Rech 2009).
4. Inter-crossing relationships (Williams 2009, Bowers-Campbell 2008, Knobel and Lankshear 2009).
5. Communication skills (Wang and Braman 2009, Shen and Eder 2009).
6. Environment-friendly behaviour (Harris and Rea 2009, Nealy 2009).

Figure 1 shows the six main positive impacts of SNs on students in Saudi Arabia. The six impacts can be further divided into smaller variables that represent each major component of positive impact of SNs. The six main advantages of SNs for students are identified from the literature review conducted previously. The concept depicted above will be used to devise an online questionnaire survey to answer the research questions.

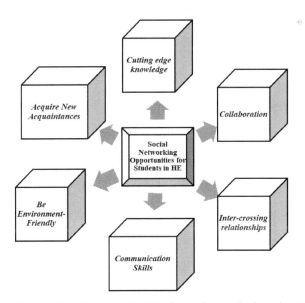

Fig. 1. Social networking opportunities for students in higher education.

The risks of social networks for students in higher education as reported in different research studies are listed below. Social networks affect

- Cognitive development (Cain 2008)
- Physical development (Bowers-Campbell 2008)
- Social development (Gibb et al. 2013)
- Security (Robelia et al. 2011)

Figure 2 shows the six main positive impacts of SNs on students in Saudi Arabia. The six impacts can be further divided into small variables that represent each major component of positive impact of SNs. Six main advantages of SNs for students are identified from the literature review conducted previously. The concept depicted above will be used to devise an online questionnaire survey to answer the research questions.

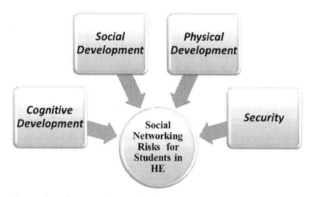

Fig. 2. Social networking risks for students in higher education.

Research Gap

The adoption of social networking in Saudi Arabia is still evolving (Alanzi et al. 2014) because Saudi Arabia focusses strongly on internet and information to reach its governance objectives and goals. Due to limited freedom of expression and SN restrictions as discussed earlier, the SN adoption process in higher education institutions is limited. Moreover, little academic research can be found regarding Saudi Arabia that addresses the role, opportunities and risks of SN in higher education (Zakaria et al. 2013). Therefore, in order to fill this research gap, a research study will be conducted to identify and analyse the real and potential risks as well as opportunities associated with the implementation of SNs in the higher education sector of Saudi Arabia. Also, this study will focus on identifying the relationships between SN and sustainability in the higher education sector. More precisely, there is a significant research gap in Saudi Arabia regarding the major real and potential risks and opportunities of social network use based on age and education. It is anticipated that this study will address this research gap.

Research Questions and Methods

This paper aims to address the research question, 'What are the major real opportunities and risks of social network use in the higher education sector in Saudi Arabia?' and 'What is the relationship between SN and sustainability in the higher education sector in Saudi Arabia?' This study will examine and investigate the effectiveness and advantages of SN, especially for students if higher education institutions implement these networks in Saudi Arabia. Therefore, research and analysis will be done on the effectiveness of social networks as a means of enhancing educational experience of students in higher education. Finally, the study will examine the method and challenges of adopting and implementing these networks in higher education in the context of Saudi Arabia.

Specifically in this study, quantitative online surveys use the positivism research philosophy. In the beginning, a critical review of literature is conducted to devise the research strategy and research questions. Then, an online survey is based on the findings of the literature review, is carried out. The literature review has enabled the researcher to gain deeper insight into the use of social networking by students in the tertiary institutions of the country. Moreover, it provided guidance regarding a feasible research approach to inductively investigate the issue. Based on these findings, online surveys have served two purposes: (1) they have enabled the researcher to answer the research questions; and (2) they have confirmed and validated the findings of the literature review in order to construct a theory. The outcomes from the survey aiming to answer the research question are objective and the participants completed questionnaire surveys.

The survey link was sent to respondents and they were requested to copy and paste the link in their browser. The link directed them to the online survey where they filled the survey electronically through the internet. All the respondents were requested to fill the survey within one week because the link to the survey had an expiry time of four months. Therefore, this time margin was required in order to finalise the research findings. Two reminders and follow-up online messages were sent to the respondents. The research participants completed the survey and responded within seven to ten days. The completed surveys allow the researcher to draw more justified and effective conclusions (Zikmund et al. 2012) about social networks in higher education. Moreover, the answers to the primary and secondary research questions would be more valid using positivism. Figure 3 explains the research methodology in pictorial form.

In this research study, the population comprises students, teachers and academic workers that are employed by universities and higher education institutions in Saudi Arabia. Due to limited time and resource constraints, it is not possible to engage the whole population in the census study. Therefore, samples of the research subjects were selected from the entire population to gather and analyse the research data.

In order to complete the research study, the researcher selected a sample of 375 students that are enrolled in a Master's programme in different Saudi Arabian universities. It should be noted that a sample of n = 30 or $n > 30$ is required to make the research findings general (Zikmund et al. 2012). In this survey, the sample size of 120 exceeds 30, indicating that the findings of this research study are generaliz and accurate. In order to select the students, convenience sampling is used. Convenience sampling is one of the convenient ways to select samples from the population and it is effective when access, costs and time resources are limited. Therefore, due to limited access, costs and time resources, the researcher preferred convenience sampling. This way, the students are easily available and

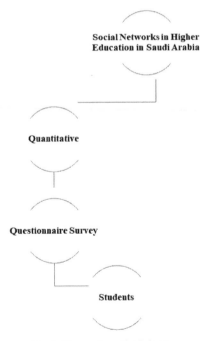

Fig. 3. Research methodology.

willing to participate in the study. However, convenience sampling is a weak sampling strategy, unlike other probability sampling techniques, such as random sampling. The primary reason for selecting this sampling is that it suits the access, time and cost constraints of the researcher and the project.

The key survey questions are divided into sections. The three sections are demographic, positive and negative aspects of SN usage. The participants and students were instructed to choose only from the stated responses while selecting only one option as an answer to the question rather than selecting multiple choices for one question. Then, the responses of the students were further quantitatively analysed and evaluated. The researcher used factor analysis and alpha to analyse the findings of this research study (Aleamoni 1976). In order to obtain e-mail addresses and social media profiles of the students, the researcher contacted the management of the universities as well as the students.

A briefing on the purpose of the study was delivered to the university management in written form and official e-mail addresses of the students were obtained from management offices with the informed consent of the universities' administrators. Secondly, the students were asked to provide WhatsApp, Facebook, Twitter and e-mail addresses of their friends and peers after explaining to them the general purpose of the study. The students were then contacted through their Whatsapp, Facebook, Twitter and e-mail addresses obtained from universities and students. The introductory message contained a written explanation of the general purpose of the study, the link to the online survey, the significance of the participants' involvement in the study and a written permission from their university to engage the students in the survey. The students were requested to follow the link to complete the survey if they agreed and were willing to participate in the survey. They were asked to complete the survey and respond within a week. Two reminders and follow-up online messages were sent to students during the week to remind them about the survey. All the respondents completed the survey within seven to ten days. Figure 4 explains the data collection process.

This research study utilises different statistical tools and tests to analyse the research findings. The analysis process is discussed in detail in the next chapter. More precisely, factor analysis and Cronbach's Alpha are used to analyse the research findings. In factor analysis, observed variables are represented in terms of common factors and also includes factors that are unique to each variable. Factor analysis enables the researcher to conduct effective scientific investigations, thereby enhancing the validity and reliability of the study (Veal and Ticehurst 2005). Therefore, in order to conduct valid and reliable research, factor analysis is selected.

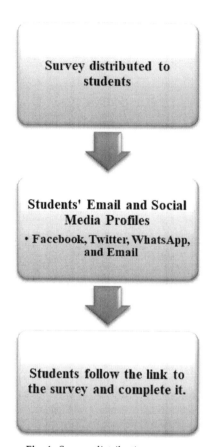

Fig. 4. Survey distribution process.

Participants

Both genders among 375 students participated in this study with 126 students completing the whole survey. Therefore, 74 per cent of the total respondents in the study were male (see Fig. 5 and Fig. 6) and 26 per cent were female (total respondents 126).

A point that should be noted is that Saudi Arabia has strong control over access to information and the internet, especially in the case of female students due to Islamic values. Islam honours women and it is the responsibility of the Islamic government to prevent the women from immoral and unethical consequences of having free access to information (Alanzi et al. 2014). Therefore, the Saudi government has a significant control of the internet, especially regarding its use by women, as was discussed in literature (Zakaria et al. 2013). Thus, the researcher received fewer responses from females because most of them were not aware of social

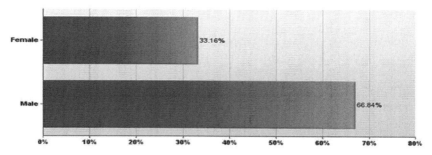

Fig. 5. Number of participants.

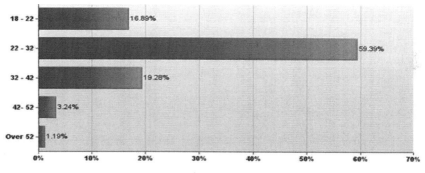

Fig. 6. Age.

networks, while others were not willing to participate in the study. Hence, another claim made in the literature was confirmed. The respondents belonged to varying area of specializations. It has been discussed in the literature review that the majority of healthcare educational institutions have implemented SNs for education purposes.

This was confirmed in this study as the researcher obtained 20 per cent of the responses only from the students specialising in some area of healthcare. Students belonged to different areas of specialisation, including accounting, business law, economics and finance, information systems, information technology, computer science, management, marketing, health sciences, humanities, science and engineering, art and design and other areas. The majority of responses were obtained from health sciences and science and engineering at 20 per cent and 17 per cent respectively. It was found in the literature review that SNs were initially implemented in the medical education and healthcare sector. Therefore, SNs have been mature in this sector and most of the students from the healthcare sector were active participants in this study.

Results

Under this section the online survey results will be examined and discussed by the researchers.

Opportunities Impact of Social Networks on Students in Higher Education

The four variables are directly related to the cutting edge knowledge component. It indicates and highlights that SNs enhance the cutting edge knowledge of the students in higher education. Among the four variables, acquiring up-to-date information is the most significant variable with the highest factor loading. It is also the new variable that is identified in the research findings. It suggests that students access significant information through SNs in comparison to other factors contributing to cutting edge knowledge. The factor with the second highest factor loading is being aware of global issues and local issues. The research findings suggest that the second significant advantage of SNs is that they enhance the students' awareness of local and global issues. The factor loadings in Table 2 show the significance of each factor as an advantage or positive impact of SNs on the students.

Collaboration being a component of positive impact of SNs comprises four different factors. The above findings reveal that peer collaboration among the students increases significantly through the use of SNs because this factor has the highest factor loading compared with other factors. Therefore, it is the new identified variable that shows that social networks increase collaboration of students with their peers. The second most significant factor is that SNs enable the students to communicate with their peers more frequently and easily (see Table 3).

The third impact of SNs on students is that it enhances their inter-crossing relationships as they are able to conduct their research studies more easily. This has the highest factor loading compared with other factors that makes it a new variable reflecting impact of SNs on relationships. It is the SN factor with the most significant impact on students. Moreover, these findings indicate that personal and professional development of students enhances significantly as they are able to study independently while they complete their studies more quickly as the respective factor loadings suggested (see Table 4).

In this study, the various factors pertaining to communication skills of the students were tested. The findings indicate that the most significant positive factor associated with student use of SNs is that it develops personal and communication skills. Moreover, their professional attitude towards study and work is enhanced. All these factors are directly related to communication skills; hence, the most significant impact of SNs is on the communication skills of the students in Saudi Arabia (see Table 5).

Environment friendliness was identified as an important impact of SNs on students in literature review. The above findings reflect that students are able to provide reliable and scalable services with SNs and the highest factor loading. It makes this variable as a new identified variable reflecting environmental friendliness caused by SNs in higher education. Therefore, it is a most significant factor among the remaining three factors or positive impacts of SNs. With a slightly lower factor loading, the second most significant factor is that students become greener in their activities by engaging in SNs in higher education (see Table 6).

The findings of the final component suggest that students form new friendship relationships more frequently and significantly through SNs compared to other factors contributing to acquiring new acquaintances. Also the other relationships are significantly lower than 0.847, indicating their low significance as an impact of SNs on students, but they are significant as they show a factor loading of more than 0.5 or level of significance (see Table 7).

It was found that the most significant positive factors affecting by social networks include acquisition of latest information (cutting edge

Table 2. Cutting edge knowledge.

Component Matrix		
	Component	New Factor
	1	
Learn new information and knowledge	.773	Acquire latest information
Gain up-to-date information	.876	
Be more aware of global issues/local issues	.839	
To remember facts/aspects of the past	.690	

Table 3. Collaboration.

Component Matrix		
	Component	New Factor
	1	
Communicate with my peers frequently	.858	Regular collaboration with peers
Collaborate with my peers frequently	.886	
Communicate with my peers from different universities	.835	
Communicate with my different communities	.809	

Table 4. Intercrossing relationships.

Component Matrix		
	Component	New Factor
	1	
Develop inter-crossing relationships with my peers (i.e., artistic talents, sport and common interests)	.610	Scrutinize research and study easily
Study independently	.848	
Overcome study stress	.719	
Complete my study more quickly	.843	
Understand and solve study problems easily	.839	
Scrutinize my research study more easily	.862	

Table 5. Communication skills.

Component Matrix		
	Component	New Factor
	1	
Develop my personal and communication skills	.881	Improve personal and communication skills
Concentrate more on my reading and writing skills	.865	
To prepare my professional attitude toward study and work	.876	

Table 6. Environment friendly.

Component Matrix		
	Component	New Factor
	1	
Be more sustainable person	.806	Sustainable and accessible activities
Provide reliable and scalable services	.882	
Become more "Greener" in my activities	.881	
Reduce carbon footprint in my activities	.841	

knowledge), regular collaboration with peers (collaboration), scrutiny of research and study easily (inter-crossing relationships), improve personal and communication skills (communication skills), sustainable and accessible activities (environment friendly), acquire new friendship relationships (acquire new acquaintances). Figure 7 presents the new factors for social networking opportunities for students in higher education.

Table 7. Acquire new acquaintances.

Component Matrix		
	Component	New Factor
	1	
Acquire new acquaintances—work related	.774	Acquire New Friendship relationship
Acquire new acquaintances—friendship relationship	.847	
Acquire new acquaintances—romance relationship	.768	
Do whatever I want, say whatever I want, and be whoever I want	.707	

Fig. 7. Social networking opportunities for students.

Risks Impact of Social Networks on Students in Higher Education

It was discussed in the literature that SNs had a negative impact on the cognitive development of students. It is indicated by the above findings that SNs significantly decrease deep thinking among students as this has the highest factor loading of 0.830. Therefore, this factor is most significant among other factors and it is also a newly identified variable relative to other variables that are related to cognitive development of students, including prevention from concentrating more on writing and reading skills that has the second highest factor loading (see Table 8).

Table 8. Cognitive development.

Component Matrix		
	Component	New Factor
	1	
Prevents me from concentrating more on writing and reading skills	.736	Decline students' deep thinking
Prevents me from remembering the fundamental knowledge and skills	.823	
Scatters my attention	.664	
Decreases my grammar and proofreading skills	.777	
Decreases my deep thinking	.830	
Distracts me easily	.668	

It was identified in the literature that SNs negatively impacted on the social development of students. In the findings below, the new variable and factor 'stresses me' reflects the highest factor loading compared with other factors. Therefore, it is the most significant negative impact of SN on the social development of students. The second most important factor is loneliness with second highest factor loading as less socially developed students do not socialize and they want to remain alone (see Table 9).

Further the literature claimed that SNs impacted the physical development of students. This component was tested with the finding below and suggested that lack of face-to-face contacts is the most significant negative factor that reduces the physical development of students. When students engage in virtual interactions, their physical development is affected. Secondly, the students are unable to complete their work on time and is the second most significant factor affecting physical development caused by use of SNs. It has the second highest factor loading (see Table 10).

The findings in Table 11 indicate that increasing security concerns have a significant negative impact caused by SNs that raise security concerns for students. It has the highest factor loading that is followed by increased privacy concerns with the second highest factor loading. The following concept map describes the newly identified variables in the study.

From the above discussion, the most significant negative factors include decline in students' deep thinking (cognitive development), stress and anxiety (social development), lack of personal interaction (physical development), and security and fears (security). It further implies that thinking capability as well as stress in students increase due to social networks when they do not find effective face-to-face interaction opportunities that impact their physical development. Also, their security

Table 9. Social development.

Component Matrix		
	Component	New Factor
	1	
Prevents me from participating in social activities	.672	
Prevents me from completing my work/study on time	.620	
Makes me sick and unhealthy	.784	Stress and Anxiety
Bores me	.785	
Stresses me	.836	
Depresses me	.808	
Makes me feel lonely	.829	
Makes me lazy	.756	
Makes me addict	.711	
Makes me more gambler	.708	
Makes me insecure to release my personal details from the theft of personal information	.724	
Makes me receive an immoral images and information from unscrupulous people and it is difficult to act against them at present	.667	

Table 10. Physical development.

Component Matrix		
	Component	New Factor
	1	
Prevents me from having face to face contact with my family	.809	Avoid personal interaction
Prevents me from having face-to-face contact with my friends	.840	
Prevents me from participating in physical activities	.797	
Prevents me from shopping in stores	.741	
Prevents me from watching television	.740	
Prevents me from reading the newspapers	.668	
Prevents me from talking on the phone/mobile	.764	
Prevents me from completing my work on time	.822	
Prevents me from completing my study on time	.752	

Table 11. Security.

Component Matrix		
	Component	New Factor
	1	
Increase privacy concerns	.948	Security and fears
Increase security concerns	.949	
Increase intellectual property concerns	.895	

concerns increase while they use social media. These new factors were found to be significant relative to other factors. Figure 8 presents the new factors for social networking risks for students in higher education, especially in Saudi Arabia.

Fig. 8. Social networking risks for students in Saudi Arabia.

Discussion and New Findings

In the above discussion, it is found that the research findings matched directly with the literature review findings as well as the research study. The research study answered the research questions. This study identified, produced and confirmed six opportunities, namely: acquire new friendship relationship; sustainable and accessible activities; improve personal and communication skills; scrutinize research and study easily; regularly collaborate with peers and acquire latest information. Moreover, the four risks are, namely decline in students' deep thinking; stress and anxiety; lack of personal interaction; security and fears associated with SN usage in Saudi Arabia.

In the research questions, four factors were used to analyse the impact of SNs on the cutting edge knowledge of students. It was found that the acquisition of up-to-date information is the most significant variable with the highest factor loading of 0.876. This indicates that SNs directly enabled students to access updated and new information and this has a positive impact on their cutting edge knowledge.

The second opportunity of SNs for students was identified to be collaboration that was further tested by dividing it into four different factors. It was found that SNs enabled collaboration among students by encouraging and enabling them to collaborate with their peers frequently. This factor had the highest factor loading of 0.886. Therefore, when collaboration with peers was enhanced, the students' learning experience and knowledge also enhanced.

It was identified in the literature review that communication skills were the third opportunity of SNs for students. Three factors were used to analyse the impact of SNs on the communication skills and it was found that SNs directly enhance the personal and communication skills of students as this factor had the highest factor loading of 0.881 compared to the other two factors.

Inter-crossing relationships as the fourth opportunity of SNs was tested by analysing six factors. It was found that SNs enabled the students to scrutinize their research study more easily and this enhanced their inter-crossing relationships. It had a factor loading of 0.862.

The fifth opportunity of SNs for students was that SNs made them environment friendly. It was tested by using four factors. It was identified that SNs enabled the students to provide reliable and scalable services that reflected their environment friendliness. Finally, acquiring new acquaintances was identified to be the last advantage of SNs. It was also tested by four factors. It was found that SNs enhanced the friendship relationships of students, thereby enabling them to acquire new friends and acquaintances. It had the highest factor loading of 0.847.

Similarly, four risks of SNs for students were tested in online questionnaire surveys. The risk of slow cognitive development was tested by using six factors. It was found that SNs directly decreased deep thinking in students and this had a negative impact on their cognitive development. It had a factor loading of 0.830.

The second risk of SN for students was that it negatively affected their social development. It was tested by 11 factors. Among the 11 factors, it was identified that SNs increased the stress of students that impacted negatively on their social development. The factor 'stresses me' had the highest factor loading of 0.836 among the 11 factors.

The third risk of SNs was that it impacted the physical development of students. Among the nine factors tested, it was found that SN reduced face-to-face contacts of students with their friends and thus had a negative impact on their physical development. It had the highest factor loading of 0.840.

Finally, security was identified as the fourth main risk of SNs for the students. It was tested by three factors and it was found that SNs increased the security concerns of students that impacted their security and privacy. It had a factor loading of 0.949.

It was concluded that the findings of the online surveys matched with the literature review findings. SN had a positive impact on students in terms of sustainability because students in Saudi Arabia were engaged in more environmentally responsible, greener and sustainable activities. Moreover, students started adopting environment-friendly behaviour by using SNs in higher education. Secondly, in the research findings, it was identified that there were six main advantages of SNs for students in higher education in Saudi Arabia that were: cutting edge knowledge, collaboration, acquiring new acquaintances, be environment-friendly, communication skills, and inter-crossing relationships. When students engaged in SNs in higher education, they benefitted from these positive aspects.

This research study corroborated the research questions and the literature review, and identified new significance by adopting social networking in higher education, especially in Saudi Arabia by understanding the actual prospective risks and opportunities of introducing social networking (SN) as a teaching and learning tool in the higher education sector in Saudi Arabia. Furthermore, both the research questions are answered by the research findings and match the literature review findings perfectly. Furthermore, the research study confirmed that use of social networking via higher education students in Saudi Arabia makes them more sustainable and provides reliable and scalable services. This also confirmed the second research question. The research study identified six opportunities and four risks for adopting social networking in higher education, especially in Saudi Arabia, and this answered the first research question.

Research Limitations

Under this section the researchers will discuss the research limitations:

1. The study is quantitative in nature and it lacks qualitative evidences, such as detailed descriptions and theoretical explanations to justify the findings and claims. The researcher could have improved the design of the study by using a mixed research method.

2. The researcher engaged 375 respondents in the study with his limited resources in limited time by using internet. The participants were engaged only in online questionnaire surveys, and the researcher did not have face-to-face contact or communication with the research participants while conducting the survey. The lack of face-to-face interaction is a significant communication barrier between the researcher and the participants. Due to lack of such interactions, it is possible that the researcher may have misunderstood, misrepresented, or misinterpreted certain survey questions because the questionnaire survey was completed by the participants.

3. Face-to-face interaction was not the only reason for misunderstanding of the survey questions. In Saudi Arabia, the native language is Arabic and it is used in daily life. Secondly, Arabic is used for official communications and education is delivered mainly in the Arabic language. Questionnaire surveys were designed in English. Therefore, due to language barriers, several participants were having difficulty in understanding and engaging in the survey. This could have impacted the research findings.

4. The response rate was too low than expected and only 126 participants completed the survey. With increased response rate, the research findings would have been more valid and reliable.

Conclusion

This research study was intended to answer two research questions regarding SNs and their impact on the students in Saudi Arabia. The first question was concerned with identifying the relationship between sustainability and SN in higher education, while the second question was concerned with identifying the major risks and opportunities associated with student use of SNs, based on age and education in the higher education sector in Saudi Arabia. The researcher conducted a critical review of the literature and identified six advantages and four risks associated with students using SNs. The opportunities were cutting edge knowledge, collaboration, acquiring new acquaintances, be environment-friendly, communication skills, and inter-crossing relationships. The risks were associated with cognitive, social and physical development, and security. Moreover, theoretical insights regarding the relationship of SN and sustainability were obtained. On the basis of the literature review, a questionnaire survey was designed and 375 students from higher education sector of Saudi Arabia were recruited for online questionnaire surveys. It was found that there was a positive relationship between sustainability and SNs in higher education.

Moreover, literature review findings were confirmed as the research findings validated that SNs had the above-stated six advantages and four disadvantages for students in higher education in Saudi Arabia.

SNs have a positive relationship with sustainability in the higher education sector. Secondly, SN creates several opportunities for students in higher education of Saudi that are: cutting edge knowledge, collaboration, acquisition of new acquaintances, environment friendliness, communication skills, and inter-crossing relationships. Also, there are certain risks that are associated with SNs including risks related to cognitive development, social and physical development, and security.

References

Alanzi, T. M., R. S. Istepanian, N. Philip and A. Sungoor. (2014). A study on perception of managing diabetes mellitus through social networking in the kingdom of Saudi Arabia. XIII Mediterranean Conference on Medical and Biological Engineering and Computing 2013 pp. 1907–1910.

Aleamoni, L. M. (1976). The relation of sample size to the number of variables in using factor analysis techniques. Educational and Psychological Measurement 36(4): 879–883.

Al-Khalifa, H. S. and R. A. Garcia. (2013). The state of social media in Saudi Arabia's higher education. International Journal of Technology and Educational Marketing (IJTEM) 3(1): 65–76.

Boon, S. and C. Sinclair. (2009). A world I don't inhabit: disquiet and identity in Second Life and Facebook. Educational Media International 46(2): 99–110.

Bowers-Campbell, J. (2008). Cyber "Pokes": Motivational antidote for developmental college readers. Journal of College Reading and Learning 39(1): 74–87.

Cain, J. (2008). Online social networking issues within academia and pharmacy education. American Journal of Pharmaceutical Education 72(1): 10.

Chauhan, K. and A. Pillai. (2013). Role of content strategy in social media brand communities: a case of higher education institutions in India. Journal of Product & Brand Management 22(1): 40–51.

Cheung, C. M., P.-Y. Chiu and M. K. Lee. (2011). Online social networks: Why do students use facebook? Computers in Human Behavior 27(4): 1337–1343.

Dabbagh, N. and A. Kitsantas. (2012). Personal learning environments, social media, and self-regulated learning: A natural formula for connecting formal and informal learning. The Internet and Higher Education 15(1): 3–8.

Das, S. K., K. Kant and N. Zhang. (2012). Handbook on Securing Cyber-Physical Critical Infrastructure. New York: Elsevier.

Deng, L. and A. H. Yuen. (2009). Blogs in higher education: Implementation and issues. TechTrends 53(3): 95.

Ellison, N. B., C. Steinfield and C. Lampe. (2007). The benefits of facebook "friends:" social capital and college student's use of online social network sites. Journal of Computer-Mediated Communication 12: 1143–1168.

Erkollar, A. and B. Oberer. (2013). Putting google+ to the test: assessing outcomes for student collaboration, engagement and success in higher education. Procedia—Social and Behavioral Sciences 83(4): 185–189.

Gibb, A., G. Haskins and I. Robertson. (2013). Leading the entrepreneurial university: meeting the entrepreneurial development needs of higher education institutions. Universities in Change - Innovation, Technology, and Knowledge Management 9–45.

Gray, R., J. Vitak, E. W. Easton and N. B. Ellison. (2013). Examining social adjustment to college in the age of social media: Factors influencing successful transitions and persistence. Computers & Education 67: 193–207.

Greenhow, C. (2011). Online social networks and learning. On the Horizon 19(1): 4–12.

Harris, A. L. and A. Rea. (2009). Web 2.0 and virtual world technologies: A growing impact on IS education. Journal of Information Systems Education 20(2): 137.

Huang, W.-H. D., D. W. Hood and S. J. Yoo. (2013). Gender divide and acceptance of collaborative Web 2.0 applications for learning in higher education. The Internet and Higher Education 16: 57–65.

Knobel, M. and C. Lankshear. (2009). Wikis, digital literacies, and professional growth. Journal of Adolescent & Adult Literacy 52(7): 631–634.

Kumar, R., J. Novak and A. Tomkins. (2010). Structure and evolution of online social networks. Link Mining: Models, Algorithms, and Applications 337–357.

Lewis, B. and D. Rush. (2013). Experience of developing Twitter-based communities of practice in higher education. Research in Learning Technology 21: 18598.

Li, L. and J. P. Pitts. (2009). Does it really matter? Using virtual office hours to enhance student-faculty interaction. Journal of Information Systems Education 20(2): 175–185.

Li, Y., H. Chen, Y. Liu and M. W. Peng. (2014). Managerial ties, organizational learning, and opportunity capture: A social capital perspective. Asia Pacific Journal of Management 31(1): 271–291.

Ma, C. and N. Au. (2014). Social media and learning enhancement among chinese hospitality and tourism students: a case study on the utilization of tencent QQ. Journal of Teaching in Travel & Tourism 14(3): 217–239.

McGuinness, M. and R. Marchand. (2014). Business continuity management in UK higher education: a case study of crisis communication in the era of social media. International Journal of Risk Assessment and Management 17(4): 291–310.

Mitra, S. and R. Padman. (2014). Engagement with social media platforms via mobile apps for improving quality of personal health management: a healthcare analytics case study. Journal of Cases in Information Technology 16(1): 73–89.

Nealy, M. J. (2009). The new rules of engagement. Diverse: Issues in Higher Education 26(3): 13.

Ras, E. and J. Rech. (2009). Using Wikis to support the Net Generation in improving knowledge acquisition in capstone projects. Journal of Systems and Software 82(4): 553–562.

Robbins-Bell, S. (2008). Higher education as virtual conversation. Educause Review 43(5): 24.

Robelia, B. A., C. Greenhow and L. Burton. (2011). Environmental learning in online social networks: adopting environmentally responsible behaviors. Environmental Education Research 17(4): 553–575.

Sancho, T. and F. de Vries. (2013). Virtual learning environments, social media and MOOCs: key elements in the conceptualisation of new scenarios in higher education: EADTU conference 2013. Open Learning: The Journal of Open, Distance and e-Learning 28(3): 166–170.

Schroeder, J. and T. J. Greenbowe. (2009). The chemistry of facebook: using social networking to create an online community for the organic chemistry. Innovate: Journal of Online Education 5(4).

Selwyn, N. (2009). Faceworking: Exploring students' education related use of "Facebook". Learning, Media and Technology 34(2): 157–174.

Shen, J. and L. B. Eder. (2009). Intentions to use virtual worlds for education. Journal of Information Systems Education 20(2): 225.

Stack, M. L. (2013). The Times Higher Education ranking product: visualising excellence through media. Globalisation, Societies and Education 11(4): 560–582.

Sturgeon, C. M. and C. Walker. (2009). Faculty on Facebook: Confirm or Deny? Annual Instructional Technology Conference. Cleveland, TN: Lee University.

Veal, A. (2005). Business Research Methods (2nd ed.). Frenchs Forest NSW 2086, Australia: Pearson Education.

Veletsianos, G., R. Kimmons and K. D. French. (2013). Instructor experiences with a social networking site in a higher education setting: expectations, frustrations, appropriation, and compartmentalization. Educational Technology Research and Development 61(2): 255–278.

Wang, Y., and J. Braman. (2009). Extending the classroom through second life. Journal of Information Systems Education 20(2): 235.

Wankel, C. (2009). Management education using social media. Organization Management Journal 6(4): 251–262.

Williams, H. (2009). Redefining teacher education programs for the 21st century. Diverse Issues in Higher Education 26(3): 25.

Zakaria, N., A. Jamal S. Bisht and C. Koppel. (2013). Embedding a learning management system into an undergraduate medical informatics course in Saudi Arabia: Lessons learned. Medicine 2.0: 2(2).

Zikmund, W. G. (2010). Business Research Methods. New York: South-Western Cengage Learning.

CHAPTER 14

Animal Social Networks Analysis

Reviews, Tools and Challenges

Ashraf Darwish,[1,]* *Aboul Ella Hassanien*[2] and
Mrutyunjaya Panda[3]

Introduction

Animal Social Networks

Social network analysis has been used in social and computer sciences for many years (Wasserman and Faust 1994). It is as of late that biologists have likewise begun to apply informal community hypothesis to better comprehend the social association of animal groups, populaces and gatherings (Croft et al. 2004). Insights of knowledge into social network structures enable us to comprehend the functioning of animal social orders, in light of the fact that the auxiliary properties of informal organizations can effectively affect the behavior and fitness of individual animals, and can likewise encourage or compel the root, choice and flow of social procedures (Krause et al. 2015). There are numerous audit papers that describe widely how informal organization examination is significant for understanding animal behavior (Hasenjager and Dugatkin 2015, Krause et al. 2015). Collaborations can be characterized as the behavior of one

[1] Faculty of Science, Helwan University, Cairo, Egypt.
[2] Faculty of Computers and Information, Cairo University, Cairo, Egypt.
[3] Department of Computer Science, Utkal University, Vani Vihar Bhubaneswar-4, Odisha, India.
Emails: aboitcairo@gamil.com; mrutyunjaya74@gmail.com
* Corresponding author: ashraf.darwish.eg@ieee.org

animal influenced by the nearness or behavior of another. Connections are hidden variables, like, nature, hereditary relatedness, or being mates or contenders. Social associations between people can be evaluated as double (introduce or not) or weighted (relative affiliation quality). There are some affiliation lists that figure weighted association quality. In animal behavior, the weighted associations are likewise regularly utilized. Contingent upon the species and social procedure of intrigue, weighted associations may be more powerful against commotion caused by irregular affiliations and examining blunders. The relative quality of affiliations may even be a key segment of an animal's informal organization since many kinds of 'social advantages' require solid stable connections.

With the advancement of the novel innovations, spatial affiliations are moderately simple to gather in bigger amounts, taking into account quick and strong informal community examinations. Above all, the spatial vicinity is for some connections an essential, prerequisite with the exemption of signaling cooperation over long separations. Spatial closeness can be characterized in a few routes, for instance, utilizing settled separation measures, closest neighbors, chain rules, co-event in a similar gathering (gambit of the gathering methodology), or co-event on an indistinguishable area in Fig. 1.

In some species, interaction networks correlate strongly with spatial (closeness) networks, yet in others, they generally do not. It remains a challenge to determine which spatial and temporal association measures that offer the best proxies for social interactions or otherwise meaningful associations. Many advanced social network analysis packages are still restricted to binary or categorical data; however, options to analyze weighted data are expected to become more common in the near future.

Fig. 1. Representation of animal social network of deer in nodes and edges.

Motivation of Studying Animal Social Networks

The essential elements of a network are 'nodes' and 'edges'. In a graphical representation of a network, each node is represented by a symbol and every interaction (of whatever sort) between two nodes is represented by a line (edge) drawn between them. In the context of a social network, each node would normally represent an individual animal (though see later in this chapter for some alternate approaches) and each edge would represent some measured social interaction or association.

Social network theory has its origin in a number of different fields of research on humans. It goes back to the work of psychologists and sociologists in the 1930s who applied elements of mathematical graph theory to human relationships and have mostly been concerned with the scenario in which each node represents a single person and each edge some interaction or relationship between two people. A great deal of progress has been made in the analysis and modeling of human social networks in the past twenty or thirty years, made possible by the advent of readily available and cheap computing power. Network theory provides a formal framework for the study of complex social relationships. Human social networks have been used to investigate a scope of points. As opposed to pondering on human social networks, the utilization of system hypothesis to consider the social association of animal gatherings or populaces is still moderately phenomenal. For instance, Fig. 2 represents the informal community for a populace of deer.

In spite of the huge number of concentrates in animal behavior literature that have gathered data on collaborations or relationship between sets of animals, not very many examinations have utilized a system way to deal with dissecting them. We trust that system hypothesis may offer an energizing strategy to examine both new and old informational indices,

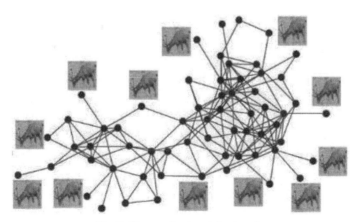

Fig. 2. Deers animal social network.

which could give bits of knowledge into the structure of animal social orders, impractical with customary techniques. By far, most of the social networks utilize one node to represent a solitary individual, and each edge represents to some type of connection or relationship between two people.

We trust that a systems approach has extraordinary potential as for quantitative examination on the animal social structure at all levels, from people to populaces. The principal point is that understanding the system structure will possibly educate us considerably more concerning the individual and populace than will data on singular qualities alone or associations between people in confinement. Informal community examination enables us to concentrate on people and inspect the impact of the system on singular behavior. Systems approach additionally can possibly light up the impact of the individual behavior on organizational structure and capacity. Most likely the primary quality of the system approach is its capability to address populace level or cross-populace level issues by working up to complex social structures from singular-level cooperation. Subsequently, the system approach overcomes any issues between the individual and populace, and this is exceptionally applicable to displaying populace-level procedures. For instance, a system approach enables us to recognize who is associated with whom in the populace or data that plans theories in the matter of who will gain from whom, or who will contaminate whom with an illness.

Flow of Information in Animal Social Networks

Social network examinations can be centered on various levels of the system: people, potential groups and the aggregate system. In investigation of animal behavior, examinations of individual-based (e.g., node) measures are generally normal. A significant number of these measures concentrate on node centrality, on how an individual itself is associated with alternate people inside the system, both straightforwardly (degree) and by implication. How social networks anticipate data moves through a populace is a key point in animal informal community investigation and can be tried by means of system-based dissemination examination. By incorporating the planning or request of data or ability obtaining with an informal organization structure, a few examinations have distinguished social transmission and even the transmission of culture. The probability of finding new fixes was biggest for the most focal people in the system. System-based dissemination investigation can be utilized to recognize properties encourage or limit social transmission. It can likewise be utilized to look if particular instruments of social transmission change with the sort of data procured or with the kind of system watched.

Communication in Animal Social Networks

Animal Communication Networks

All animals convey somehow, and the idea of a correspondence framework reflects and detects animals' social relations. Signals can be utilized at the short and long range, utilizing one or different signaling modalities, and are entered in helpful and focused relations, mate decision, rummaging behavior, predator shirking, or asset protection. Signals can reflect past activities, foresee future activities, and give key data about signaler's attributes, for example, quality, formative foundation, inspiration, or identity (Hasenjager and Dugatkin 2015). Signals are a necessary piece of social behavior. The more particular term 'correspondence organize' was at first settled in a fundamental paper by (Ilany and Akçay 2016), who characterized correspondence including signaling collaborations inside which spies can identify and utilize asymmetries in associations as a wellspring of data to evaluate relative contrasts among conspecifics. The key idea here is that signaling interactions are often asymmetric, with one signaler utilizing a signal uniquely in contrast to its adversary, regarding timing or basic examples, and that these topsy-turvy connections can give data that isn't accessible by taking care of each of the signals independently. Transmission can likewise be utilized to look at the particular components of social transmission differ with the kind of data procured or with the sort of system that is watched. Figure 3 explores development research in the area of animal social networks with computer social networks. The statistics in this figure are taken from the Web of Knowledge by using social and animal social networks keywords.

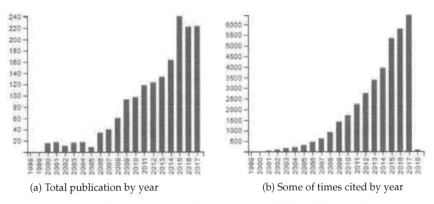

(a) Total publication by year   (b) Some of times cited by year

Fig. 3. Distribution of animal and social networks since 2000 to 2017 according to web of science with different keywords.

Combination of Animal Communication and Social Networks

Many signals in animals are obvious and contact different people, thereby associating them with the data stream. Since signals have developed to impact others' choices, they undoubtedly are a vital piece of informal for organizations in a more extensive sense. In this way, correspondence is at any rate as pertinent as vicinity is in understanding social relations among people and gatherings (Honarmand et al. 2015, McGregor and Dabelsteen 1996, Snijders et al. 2015). As noted before, correspondence systems, in spite of the comparative wording, have concentrated essentially on signaling cooperation of single direct observers or spies, and on the sort of data transmitted. The social network approach, interestingly, has concentrated on more vast-scale affiliation investigations, including roundabout associations and their results over an extended number of people, yet regularly not including data on the real nearness of connections and their tendency. To completely evaluate correspondence in social networks, a traditional correspondence arrange approach and other individual signaling behavior, should be coordinated inside the informal community system. There are various signaling qualities to be considered when we need to comprehend the part of correspondence over a bigger society and its connections with informal organizations. In the accompanying area, we consider in more detail how signaling attributes reflect and influence (closeness-based) informal organizations.

Figure 4 depicts the calculated diagram on social signals in animal social vicinity systems. Social signals can give important social data about the signaler (personality, inspiration, attributes, setting, area and so on). Be that as it may, the unwavering quality and adequacy of these signals may

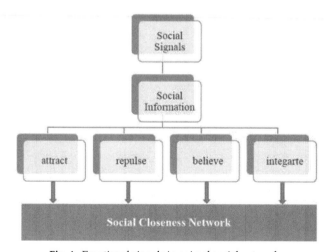

Fig. 4. Emotional signals in animal social network.

be influenced by a signal debasement or ecological clamor. The social data can impact fascination and repugnance of immediate and backhanded beneficiaries. The social data conveyed by the signals can likewise advance or smother coordination inside a social gathering or impact the social notoriety of the signaler. By influencing incorporation, notoriety and nearness, social signals subsequently influence informal organizations. In the meantime, informal communities (social communications) impact the improvement and utilization of social signals (Snijders and Naguib 2017).

In this unique situation, we additionally consider the part of independently fluctuating signaling techniques and how they may impact non-random spatial relationships among people of a population.

Animal Social Network Properties

Animals are commonly not inactive on-screen characters reacting to the social condition, yet rather frequently impact the social condition themselves. A few animals are disproportionally compelling, which turn people. There is a reasonable incentive to foreseeing and knowing the people who are basic for assembling steadiness. For instance, the expulsion of people from the wild for human utilization, amusement, or natural life trafficking can possibly destabilize the social gatherings and conceivably even populaces. The impact of the particular evacuation of particular people, or the pulverization of areas that encourage social affiliations can be reproduced. Reproduction would then be able to be to create forecasts about the impact of genuine evacuation. Likewise with evacuations, one can think about the potential impact of presenting new animals tentatively or utilizing virtual re-enactments. Experiences on the part of rotating people and the social availability of animal populaces when all is said can be indispensable for anticipating the strength of populaces to unsettling influence. Informal organization examinations can see how social discontinuity and unsteadiness can be averted; yet, perhaps at the same time, how social dependability can be reestablished.

Signals and Vocal Interaction as Social Network Components

To date, joining signals and signaling collaborations in nearness systems has been extraordinary, with a couple of remarkable cases. These cases are based mostly on visual and material signals and not sound-related signals, despite the fact that sound-related signals can interface with people over long separations and give signals about hereditary bonds and social relations. Also, vocal cooperation designs are very much contemplated and can give vital data about the course and nature of connections among people. Surely, the utilization of vocal signals for making derivations about social structures has been particularly significant in animal species that are hard to watch or track spatially in the wild, for example, marine warm-blooded animals (Garland et al. 2011). With current improvements in mechanisation, recording and sound examination systems, including such individual signal data, will probably get more extensive consideration later on. Animal social networks are influenced by the capacity of animals to recognize classes of people and regularly among singular characters. Class-level acknowledgment and individual acknowledgment have been recorded for a wide assortment of animal species. In specific species, singular signals can be retained for a drawn-out stretch of time. Male hooded songbirds, for example, still recalled singular neighbor melodies while coming back from relocation, following eight months of detachment (Godard 1991). Signals that convey data on singular personality (character signals) can help in doling out specific signals and signaling communications to particular people. There is an assortment of vocal characteristics and structures that permit animals (and scientists) to remove singular personalities, including peeps, codas, shrieks and melodies. The fish evoke an altogether higher number of 'plunges' (jumping shows) when hearing alien trills, contrasted with when hearing occupant twitters. In addition, when the trills of two closest neighbors are exchanged, a higher number of plunges are seen when tweets are communicated from the right domains. Signals give an extensive variety of data about inspiration and quality. This incorporates data about animals' physiological condition, mating status, availability to safeguard assets, inspiration to heighten challenges or identity. Such data about individual animals can be reflected in signals themselves, or in the way signals are utilized as a part of specific setting or social association. Other than conveying individual-related data, motions in signaling connections can convey pertinent data about the nearness and nature of entombing singular social bonds. The signals on which conspecifics listen stealthily and react to signaling connections, the concentration of exemplary correspondence organize ponders, particularly give data on relative phenotypic contrasts and social connections. Similarly, in non-human animals, signal connections can convey data on whether to think about a relationship as agonistic or affiliative. However, given that the data esteem and significance of many signals differ crosswise over settings and species, the mind is required when data from certain signal connections is utilized to draw inductions about social relations. Applicable data about connections among people can be encoded not simply in the direct link between singular signal communications, yet additionally in the likeness of certain signal qualities, as in lingos. The variety of people in the similitude of signals has been distinguished in different taxa and has uncovered spatial, natural and, furthermore, social relations among people.

Social Attraction Based on Signals

As depicted in the previous section, signaling behavior can effectively affect informal community structure. Social animals are pulled in to conspecifics for a wide assortment of reasons, including access to nourishment, access to mates, security from predators and social thermoregulation. Conspecific fascination is frequently interceded by signals and numerous species utilize signals that are effectively transmitted with the capacity to pull in or discover others. Perching people react to social calls, and flying bats, in reality, enter perches in which perching people have reacted. Flying bats likewise separate between reactions from natural versus new perch mates, while perching bats don't segregate calls of recognizable and new conspecifics. Social calls, in this species, fill in as imperative factors in molding social closeness. As a rule, bats demonstrate a high variety of social structures and this assorted variety is probably directed partially by social calls. Contingent upon which case of signaling is utilized to pull in or repulse conspecifics, singular animals may accordingly be utilizing diverse signal and reaction techniques to upgrade or decrease the odds of social experiences. The corruption of signals can influence different people in an informal organization in various ways. Most clearly, separation and living space subordinate signal corruption, restricting data accessible to beneficiaries at a separation. A few examinations have demonstrated that debased signals evoke weaker reactions by others, either in light of the fact that corrupted signals need specificity in data or on the grounds that they are seen as originating from more remote distance, and in this manner, don't require a solid or dire reaction. Signaling separation may likewise influence essential choice procedures as signals recognized from a separation may go about as a general attractor or repellent, while signals communicated at short proximity may create more particular decisions in light of signal segregation and acknowledgment.

Social and Animal Network Analysis

Analytic Methods and Algorithms

In this area, we investigate the similarity techniques for informal community examination and arrange investigation keeping in mind the end goal to comprehend the structure of the animal system and the passionate relationship and closeness between various animals with various kinds. Animal social network investigation is a term that envelops distinct and structure-based examination, like an auxiliary investigation. It is vital on the off chance that one needs to comprehend the structure of the system in order to pick up bits of knowledge about how the system functions and settle on choices upon it by either analyzing node/

connect attributes (e.g., centrality) or by looking at measurements of the entire system attachment (e.g., thickness) (Hansen et al. 2010, Kolaczyk and Csárdi 2014). Contrasting systems, following changes in a system after some time, uncovering groups and critical nodes, and deciding the relative position of people and bunches inside a system are some of its basic strategies. These include either a static or dynamic investigation; the former presumes that an informal organization changes steadily after some time and examination on the whole system should be possible in bunch mode. Alternately, dynamic investigation, which is more multifaceted, includes spilling information that is advancing in time at a high rate. Dynamic examination is frequently in the zone of cooperation between elements though static investigation manages properties like availability, thickness, degree, breadth and geodesic separation. This segment gives a portrayal of the fundamental techniques and calculations identified with social network examination which are the territories that are presently used to break down and process data from social-based sources.

Network Analytics

Generally, animals live in an associated world in which correspondence systems are interwoven with everyday life. In animal social networks, people collaborate with each other and give data on their inclinations and connections, and these networks have turned out to be vital devices for aggregate knowledge extraction. These associated systems utilize charts and system systematic strategies can be connected to them for separating valuable information. Diagrams are structures framed by an arrangement of vertices (likewise called nodes) and an arrangement of edges, which are associations between sets of vertices. The data separated from the animal social network refers to a diagram in which the vertices or nodes speak to the animals (clients in informal organizations) and the edges speak to the connections among them, for example, a re-tweet of a message or a most loved stamp in Twitter in PC informal communities.

Link Analysis

Link-analysis is utilized to assess associations between nodes. Understanding the arrangement and advancement of such associations in informal communities requires longitudinal information on both social connections and shared affiliations. Connection mining is normally connected with content mining and can be utilized for arrangement, expectation and bunching or affiliation of rules disclosure. It is relevant in community proposal frameworks to recognize a gathering of companions with comparable interests. PageRank, which utilizes heuristic tenets, is the renowned connection examination utilized by Google to arrange web index. Of late, Google has reported the supplanting of PageRank with a more productive hunt calculation called BrainRank which depends on Profound Learning Systems. PageRank and HITS calculations are additionally utilized as a part of impact examination (Alhajj and Rokne 2014, Online). Google examination utilizes investigation capacities to decide what number of guests achieve a specific goal page. Graphical models are effective apparatuses that can be utilized to model and gauge complex measurable conditions among factors. Associated segment is another fascinating factual metric, which takes into account the investigation of data scattering in a social network. An associated part in a chart is alluded to as an arrangement of nodes and edges where a way exists between any two nodes in the set.

Community Detection Algorithms

The community identification issue in complex systems has been the subject of many investigations in the field of information mining and social network examination. A group in a chart which can be effectively mapped into a group. In spite of the vagueness of the group definition, various strategies have been utilized for recognizing groups. Arbitrary strolls, phantom bunching, measured quality augmentation and factual mechanics have all been connected to recognizing groups (Santo 2010). These calculations are ordinarily in view of the topology data from the chart or system. Identified with diagram availability, each group ought to be associated; that is, there ought to be numerous ways that interface with each match of vertices inside the bunch.

Groups constitute an imperative part of systems that are critical for both investigating a system and foreseeing associations that are not yet watched. Group discovery is basically an information-bunching issue, where the objective is to allocate every node to a group or group in some sensible way. One approach to characterize a group is by structure, e.g., groups as clubs. Faction or finish diagram is where each node is associated with each other node in the inner circle. Another example of seeing someone is to find how much a performing artist exists in a firmly bound gathering or on the off chance that he has associations outside their own particular gathering. To investigate such a thought of system grouping, dyad and set of three registration have been utilized. A dyad is a sub-chart that speaks to a couple of on-screen characters and the conceivable edges between them while a set of three comprises of three nodes and the conceivable edges among them.

A standout amongst the most understood calculations for group discovery was proposed by Girvan and Newman (Girvan and Newman 2002). This strategy utilizes another closeness measure called 'edge between' in view of the quantity of the briefest ways between all vertex sets. The proposed calculation depends on recognizing the edges that lie amongst groups and their progressive expulsion, accomplishing the separation of the groups. The primary disservice of this calculation is its high computational many-sided quality with huge systems as in animal social network.

Arbitrary strolls can likewise be helpful for discovering groups. On the off chance that a chart has a solid group structure, an irregular walker invests a long energy inside a group due to the high thickness of inner edges and the ensuing number of ways that could be taken after. Creators in (Zhou and Lipowsky 2004) in view of the way that walkers move especially towards vertices that offer countless, defined vicinity record that demonstrates how to shut a couple of vertices to all different vertices. Groups are identified with a strategy called NetWalk, which is an agglomerative progressive bunching technique by which the similitude between vertices is communicated by their closeness. Some of these strategies are centered on finding disconnected groups. The system is apportioned into thick districts in which nodes have a bigger number of associations with each other than to whatever is left of the system, yet it is fascinating that in a few areas, a vertex could have a place with a few groups. For example, it is outstanding that animals join in an informal organization for common participation in different groups. Creators in (Xie et al. 2013) assessed the best in class covering group location calculations. This work saw that for low covering thickness systems, SLPA, OSLOM, Diversion and COPRA offer better execution. For systems with high covering thickness and high covering decent variety, both SLPA and Diversion give moderately stable execution. In any case, the test recommended that the identification of such systems is still not yet completely settled.

Information Fusion for Animal Social Networks

The animal informal organization of different kinds should be combined for giving proprietors better administration. This combination should be possible in various ways and influence various innovations, strategies and even research zones. Two of these conceivable territories are ontologies and social networks.

- **Ontology-based fusion:** Semantic heterogeneity is an imperative issue on data combination. Informal communities have characteristically extraordinary semantics from different kinds of system. Ontologies are misused from different social networks, and all the more essential, semantic correspondences are obtained by philosophy coordinating techniques.

- **Social network integration:** Next issue is the way to coordinate the appropriated social networks, for example, animal informal communities. The cloud-based stage has been connected to construct programming framework where social network data can be shared and traded (Caton et al. 2014).

Animal Social Network Analysis Tools

Graph theory is the noticeable approach in social network examination and diagram mining devices are critical in researching social structures, both logically and outwardly. Diagram databases, for example, Neo4j, graphical models, are profound learning and chart mining instruments. Systems are being created so as to proficiently deal with the need for information extraction from arranged information. Two awesome constraints in regards to animal informal community examination and information volume are (i) the confined number of extricated information from social networks in view of the restricted API exchanges and (ii) the trouble to process the information, which is beyond a specific system estimate, with chart measurements and information representations.

Challenges and Open Problems

Data Collection

To mine the datasets on animal social networks, we require preparing of perplexing and vast informational collections of animals to comprehend their passionate behavior, yet researchers and PC authorities confront the extra test of producing appropriate information in any case, by sending labels on subjects and consequently recouping tag-created data. In utilization of sciences, information is put away in the inward memory of animal-borne lumberjacks and later recovered through download from recouped gadgets or uplink transmission to recipients.

Physical Proximity

One of the key difficulties of any investigation of animal social dynamics is to outline patterns, that is, the physical closeness of at least two individuals. Bio-logging or biotelemetry frameworks offer two fundamental applied ways to deal with accomplishing this: experiences can either be recorded 'straightforwardly', with innovation that empowers animal-to-animal information trade (one snippet of data delivers an experience record), or 'by implication', with innovation that outlines the spatiotemporal positions and developments of individual animals (two snippets of data are consolidated to deliver an experience record).

Reality Mining

Reality mining empowers specialists to examine the social behavior of relatively whole human populaces in uncommon detail and with excellent spatiotemporal determination (Eagle and Pentland 2006, Eagle et al. 2009, Mitchell 2009). The example sizes accomplished by best in class human reality-mining contemplates are amazing. For instance, over 95 per cent of the human occupants of most Western nations convey and utilize cell-phones, Facebook now has roughly 900 million clients around the world and more than 400 million messages are posted on Twitter each day. With the appearance and expanding refinement of a scope of scaled down advances, scholars will soon have the capacity to copy this intense reality-mining approach in their investigations of animal social behavior and ecology, recording field informational collections of extraordinary size and quality. This will logically move the concentration from the difficulties of the information age to issues concerning information administration and examination. Reality mining ought to incorporate machine-based information accumulation, investigation and demonstrating, and the degree and structure of our audit mirror this mix of various arrangements of approaches (Krause et al. 2013).

Indirect Connection for Transferring Information

Indirect connections may likewise be imperative for the exchange of data and social practices and to the support of collaboration inside animal bunches (Bode et al. 2011, Nowak and Sigmund 2005). Aberrant associations may likewise assume a part in the transmission of sickness (Hamede et al. 2009, VanderWaal et al. 2014, Weber et al. 2013) and may impact how people react physiologically to the social condition (Brent et al. 2011). There are likewise thriving zones of research into subjects that are thoughtfully needy upon multi-specialist social associations, for example, the effect of the metagenome (i.e., the genomes of others) on social development, fellowship and quality articulation (Christakis and Fowler 2014, Slavich and Cole 2013). What's more, different regions of animal behavior research could profit by looking past dyadic affiliations, including investigations of mate decision, hybridization, and identity (Krause et al. 2010, Sih et al. 2009, Wilson and Krause 2015). However, in spite of impressive advance in our comprehension of backhanded associations in animal social behavior, a large number inquiries stay unanswered. For instance, how much data do animals host with respect to the connections of third gatherings and how frequently do they utilize this data to control, change or impact those connections? By and large, the inquiry with respect to whether circuitous associations are imperative to the lives of social animals stays open.

Data Mining Algorithms

One of the present fundamental difficulties in information mining identified with enormous information in animal social networks issues is to discover satisfactory ways to deal with breaking down of monstrous measures of online information. Since grouping techniques require past naming, these strategies likewise require extraordinary exertion for continuous investigation. Nonetheless, on the grounds that unsupervised strategies needn't bother with this past procedure, bunching has turned into a promising field for continuous examination with the information which originates from animal social network sources. At the point when information streams are examined, it is essential to consider the examination objective so as to decide the best kind of calculation to be utilized. Another age of online calculations (Fiat 1998) and streaming (Crammer and Singer 2003) is as of now being created so as to oversee animal social huge information challenges, and these calculations require high adaptability in both memory utilization (Menéndez et al. 2013) and time calculation.

Data Fusion and Data Visualization

Information combination and information representation are two critical difficulties in animal social information. Albeit both are seriously examined as to vast, disseminated, heterogeneous, and gushing information combination (Kumar et al. 2003) and information representation and examination (Keim et al. 2008). The current, quick advancement of animal social network sources with huge information advances make some especially fascinating difficulties, for example, getting solid techniques for intertwining different highlights of animals in the animal informal communities. The open issues and difficulties identified with visual investigation, particularly with the ability to gather and store new information, are quickly expanding in number, including the capacity to break down these information volumes (Wong and Thomas 2004), to record information about the development of animals at an extensive scale (Andrienko et al. 2007), to examine spatial-fleeting information and take care of spatial-transient issues in animal social networks (Andrienko et al. 2010), among others.

Concluding Remarks and Future Trends

Lately, the exploration field of animal social networks and animal correspondence systems have become key controls for understanding animal correspondence and social elements. However, to date, these controls remain inadequately incorporated. In this survey, we have featured numerous routes by which correspondence and social networks are intrinsically connected with social signs reflecting and influencing informal community segments. There are many examination bearings in animal social behavior that would profit by the reconciliation of social signs into animal social network investigation, a few of which were talked about through this survey. We are persuaded that both the field of animal social networks and the field of animal correspondence systems can profit by each other and expand upon each other's work.

The capacity to gather information simultaneously on numerous individuals has been a key imperative, albeit novel innovation which would now be able to give an abundance of information on numerous individuals acting at the same time, incorporating into their signaling connections. Utilization of automatized animal following frameworks incorporate, sensor enacted varying media playback and recording frameworks and automatized sound accounts, animals were uncovered to react to signals crosswise over long separations, prompting general changes inside the bigger neighborhood or society. Advances in innovation for information gathering, combined with novel insights of knowledge into animal correspondence and social structure, ought to permit advance coordination of social signals within the bigger social network system.

As of late, the exploration field of animal social networks and animal correspondence systems have formed into key controls for understanding animal correspondence and social elements. However, these controls remain inadequately coordinated. In this survey, we have featured numerous courses by which correspondence and informal organizations are inalienably connected, with social signs reflecting and influencing social network parts.

There are many examination bearings in animal social behavior that would profit by the joining of social signs into animal social network investigation, a few of which were talked about or said all through this audit. We are persuaded that both the field of animal informal communities and the field of animal correspondence systems can profit by each other and expand upon each other's work.

References

Alhajj, R. and J. Rokne (eds.). (2014). Encyclopedia of Social Network Analysis and Mining. New York, NY: Springer New York.

Andrienko, G., N. Andrienko and S. Wrobel. (2007). Visual analytics tools for analysis of movement data. ACMSIGKDD Explor. News 1.9(2): 38–46.

Andrienko, G., N. Andrienko, U. Demsar, D. Dransch, J. Dykes, S.I. Fabrikant, M. Jern, M.-J. Kraak, H. Schumann and C. Tominski. (2010). Space, time and visual analytics. Int. J. Geogr. Inf. Sci. 24(10): 1577–1600.

Bode, N. F., A. J. Wood and D. Franks. (2011). Social networks and models for collective motion in animals. Behavioral Ecology and Sociobiology 65: 117e130. http://dx.doi.org/10.1007/s00265-010-1111-0.

Brent, L. J. N., J. Lehmann and G. Ramos-Fern_andez. (2011). Social network analysis in the study of nonhuman primates: a historical perspective. American Journal of Primatology 73: 720e730. http://dx.doi.org/10.1002/ajp.20949.

Caton, S., C. Haas, K. Chard, K. Bubendorfer and O. F. Rana. (2014). A social compute cloud: allocating and sharing infrastructure resources via social networks. IEEE Trans. Serv. Comput. 7(3): 359–372.

Christakis, N. A. and J. H. Fowler. (2014). Friendship and natural selection. Proceedings of the National Academy of Sciences of the United States of America 111: 10796e10801. http://dx.doi.org/10.1073/pnas.1400825111.

Crammer, K. and Y. Singer. (2003). Ultra conservative online algorithms for multiclass problems. J. Mach. Learn. Res. 3: 951–991.

Croft, D. P., J. Krause and R. James. (2004). Social networks in the Guppy (Poecilia reticulata). *In*: Proceedings of the Royal Society B: Biological Sciences 271: S516eS519.

Eagle, N. and A. S. Pentland. (2006). Reality mining: sensing complex social systems. Pers. Ubiqui. Comput. 10: 255–268.

Eagle, N. et al. (2009). Inferring friendship network structure by using mobile phone data. Proc. Natl. Acad. Sci. U.S.A. 106: 15274–15278.

Fiat, A. (1998). Online Algorithms: The state of the art. *In*: Fiat, A. and G. J. Woeginge (eds.). Lecture Notes in Computer Science 1442.

Garland, E. C., A. W. Goldizen, M. L. Rekdahl, R. Constantine, C. Garrigue, N. D. Hauser and M. J. Noad. (2011). Dynamic horizontal cultural transmission of humpback whale song at the ocean basin scale. Current Biology 21: 687e691.

Girvan, M. and M. E. J. Newman. (2002). Community structure in social and biological networks. Proc. Natl. Acad. Sci. 99(12): 7821–7826.

Godard, R. (1991). Long-term memory of individual neighbors in a migratory songbird. Nature 350: 228e229.

Hamede, R. K., J. Bashford, H. McCallum and M. Jones. (2009). Contact networks in a wild Tasmanian devil (Sarcophilus harrisii) population: using social network analysis to reveal seasonal variability in social behaviour and its implications for transmission of devil facial tumour disease. Ecology Letters 12: 1147e1157. http://dx.doi.org/10.1111/j.1461-0248.2009.01370.x.

Hansen, D., B. Shneiderman and M. A. Smith. (2010). Analyzing Social Media Networks withNodeXL: Insights from a Connected World. Morgan Kaufmann.

Hasenjager, M. J. and L. A. Dugatkin. (2015). Social network analysis in behavioral ecology. Advances in the Study of Behavior 47: 39e114.

Honarmand, M., K. Riebel and M. Naguib. (2015). Nutrition and peer group composition in early adolescence: Impacts on the male song and female preference in zebra finches. Animal Behaviour 107: 147e158.

Ilany, A. and E. Akçay. (2016). Personality and social networks: A generative model approach. Integrative and Comparative Biology 56: 1197e1205.

Jens Krause, Stefan Krause, Robert Arlinghaus, Ioannis Psorakis, Stephen Robert and Christian Rutz. (2013). Reality mining of animal social systems. Trends in Ecology & Evolution.

Keim, D., G. Andrienko, J.-D. Fekete, C. Görg, J. Kohlhammer and G. Melançon. (2008). Visual Analytics: Definition, Process, and Challenges, Springer.

Kolaczyk, E. D. and G. Csárdi. (2014). Statistical Analysis of Network Data with R. Springer.

Krause, J., R. James and D. P. Croft. (2010). Personality in the context of social networks. Philosophical Transactions of the Royal Society B: Biological Sciences 365: 4099e4106. http://dx.doi.org/10.1098/rstb.2010.0216.

Krause, J., R. James, D. W. Franks and D. P. Croft. (2015). Animal Social Networks. Oxford, UK: Oxford University Press.

Kumar, R., M. Wolenetz, B. A. garwalla, J. Shin, P. Hutto, A. Paul and U. Ramachandran. (2003). Dfuse: a framework for distributed data fusion. pp. 114–125. *In*: Proceedings of the 1st International Conference on Embedded Networked Sensor Systems, ACM.

Lysanne Snijders and Marc Naguib. (March 2017). Communication in animal social networks: a missing link? Advances in the Study of Behavior. DOI: 10.1016/bs.asb.2017.02.004.

McGregor, P. K. and T. Dabelsteen. (1996). Communication networks. pp. 409e425. *In*: Kroodsma, D. E. and E. H. Miller (eds.). Ecology and Evolution of Acoustic Communication in Birds. Ithaca, New York, USA: Cornell University Press.

Menéndez, H. D., D. F. Barrero and D. Camacho. (2013). A multi-objective genetic graph-based clustering algorithm with memory optimization. pp. 3174–3181. *In*: Proceedings of IEEE Congress on Evolutionary Computation (CEC), IEEE.

Mitchell, T. M. (2009). Mining our reality. Science 326: 1644–1645.

Nowak, M. A. and K. Sigmund. (2005). Evolution of indirect reciprocity. Nature 437: 1291e1298. http://dx.doi.org/10.1038/nature04131.

Querying Social Media with NodeXL. [Online]. Available: http://scalar.usc.edu/works/querying-social-media-with-nodexl/what-is-social-media?path=index. [Accessed: 30-Apr-2016].

Santo, F. (2010). Community detection in graphs. Phys. Rep. 486(3-5): 75–174, doi:10.1016/j.physrep.2009.11.002.

Sih, A., S. F. Hanser and K. A. McHugh. (2009). Social network theory: new insights and issues for behavioral ecologists. Behavioral Ecology and Sociobiology 63: 975e988.

Slavich, G. M. and S. W. Cole. (2013). The emerging field of human social genomics. Clinical Psychological Science 1: 331e348. http://dx.doi.org/10.1177/2167702613478594.

Snijders, L., J. van der Eijk, E. P. van Rooij, P. de Goede, K. van Oers and M. Naguib. (2015). Song trait similarity in great tits varies with social structure. PLoS One 10: e0116881.

VanderWaal, K. L., E. R. Atwill, L. A. Isbell and B. McCowan. (2014). Quantifying microbe transmission networks for wild and domestic ungulates in Kenya. Biological Conservation 169: 136e146. http://dx.doi.org/10.1016/j.biocon.2013.11.008.

Wasserman, S. and K. Faust. (1994). Social Network Analysis: Methods and Applications (Vol. 8). Cambridge, UK: Cambridge University Press.

Weber, N., S. P. Carter, S. R. X. Dall, R. J. Delahay, J. L. McDonald, S. Bearhop et al. (2013). Badger social networks correlate with tuberculosis infection. Current Biology 23: R915eR916. http://dx.doi.org/10.1016/j.cub.2013.09.011.

Wilson, A. D. M. and J. Krause. (2015). Personality and social network analysis in animals. pp. 53e60. *In*: Krause, J., R. James, D. W. Franks and D. P. Crof (eds.). Animal Social Networks. Oxford, U.K.: Oxford University Press.

Wong, P. C. and J. Thomas. (2004). Visual analytics. IEEE Comput. Gr. Appl. (5): 20–21.

Xie, J., S. Kelley and B. K. Szymanski. (2013). Overlapping community detection in networks: the state-of-the-art and comparative study. ACM Comput. Surv. (CSUR) 45(4): 43.

Zhou, H. and R. Lipowsky. (2004). Network Brownian motion: a new method to measure vertex-vertex proximity and to identify communities and subcommunities. pp. 1062–1069. *In*: Computational Science-ICCS 2004, Springer.

Index

Printed and bound by CPI Group (UK) Ltd, Croydon, CR0 4YY

18/10/2024

01776242-0005